Acid-Base and Blood Gas Regulation

Acid-Base and Blood Gas Regulation

For Medical Students before and after Graduation

GILES F. FILLEY, M.D.

Professor of Medicine, Division of Pulmonary Diseases, Department of Medicine, University of Colorado School of Medicine. Pulmonary Physiologist, The Webb-Waring Institute for Medical Research, Denver, Colorado

LEA & FEBIGER · PHILADELPHIA
1971

ISBN 0-8121-0272-X

Library of Congress Catalog Card Number 74-135681

Published in Great Britain by Henry Kimpton, London

Printed in the United States of America

to

WILLIAM MANSFIELD CLARK

Preface

∙∙∙

"For the sake of persons of different types, scientific truth should be presented in different forms and should be regarded as equally scientific whether it appears in the robust form and vivid coloring of a physical illustration, or in the tenuity and paleness of a symbolical expression."   JAMES CLERK MAXWELL

∙∙∙

NUMBERS—numerical values for P_{CO_2}, $[HCO_3^-]$, pH and P_{O_2}—are now being used in the diagnosis and treatment of acid-base disturbances in medicine, surgery, pediatrics and other specialties, yet tradition still forces us to describe these disturbances in the ambiguous terminology which has beclouded this subject for half a century.

Why not describe acid-base disturbances by means of equations and numbers only? It is certainly more informative to state that the arterial pH is 7.30 than to say that it is low. Both of these statements are more definite than saying that the blood is more acid than normal. Traditionally, however, special words with Greek endings (like acidosis and acidemia) have been used to describe conditions indicated by quantitative measurements. Such qualitative description—if unambiguous—seems to be as necessary as measurement to the understanding.

In this book I shall attempt to present clinical acid-base physiology without depending on the ambiguous words of tradition. When certain of these words are used they will be marked, at least on their first appearance, by ⟨ ⟩—a signal to beware of the contagious confusion generated by them and a suggestion to look them up in Appendix I. Some of these words may eventually die of their disease, being incurably ambiguous unless accompanied by an adjective, e.g. ⟨acidosis⟩, and will be so marked throughout the book. Others, like ⟨base⟩, are probably curable and the mark of their taint will be dropped once they have been explained in the text.

Another difference from the usual approach to acid-base chemistry may be mentioned. Chemists interested in stoichometric titrations seldom work with the inconveniently volatile P_{CO_2}-bicarbonate system in the laboratory;

when teaching they therefore tend to illustrate the concept of buffering with titration curves showing how pH is determined in closed systems with conveniently non-volatile agents at equilibrium states mostly irrelevant to the behavior of dynamic biologic systems. The symbol H^+ is often omitted from chemical formulations thus concealing the central process of proton-transfer and the way potentially ionizable hydrogen is disposed of. This book starts with qualitative descriptions of the movement of acid-base materials through the body. The relationships between these materials are then displayed in a formulation built around the central concept of H^+ transfer. Finally it is shown how the normal lung and kidney cooperate to dispose of acid and alkali in health and disease and how therapeutic measures can aid or interfere with these processes.

The subject of oxygen transport will be treated along with acid-base regulation since modern measurements of blood gas pressures have shown the clinical relevance of arterial oxygenation to many acid-base problems. The aim will be to infuse medical significance into the expressions P_{CO_2}, $[HCO_3^-]$, pH and P_{O_2} since these are the four most important measurements needed to recognize and understand acid-base disturbances. The measurements can only become clinically useful, however, if they lose the tenuity and paleness of their purely quantitative aspects and take on the robust form of physiologic variables which the body regulates for its survival.

For reviewing parts of the manuscript, I wish to thank Drs. D. G. Ashbaugh, H. R. Beresford, D. B. Bigelow, J. O. Broughton, S. M. Cain, D. G. Carroll, D. C. Char, W. R. Frisell, J. L. Gamble, Jr., P. B. Oliva, J. R. Pappenheimer, T. L. Petty, J. B. Posner, E. B. Reeve, J. T. Reeves, S. A. Schneck, J. W. Severinghaus, D. H. Simmons, and J. V. Weil. I am especially grateful to Drs. Roger G. Bates, E. B. Brown, Jr., and A. Gorman Hills for their criticisms, to Miss Gladys Dart and Mrs. Ruth S. Roberts for preparing the figures and the text and to my wife for her patience.

Denver, Colorado   GILES F. FILLEY

Contents

Introduction

....

"There seems now to be no doubt that it is the diffusion pressures, and not the mere concentration of substances in the body, that are of physiological importance." J. S. HALDANE (1922)[1]

....

ACID-BASE and blood gas metabolism is a highly integrated process in which certain variables, chiefly Pco_2, $[HCO_3^-]$, pH and Po_2, are controlled. To achieve understanding of this process requires an analysis of how each component of the regulatory system (e.g. the lung) controls each variable (e.g. Pco_2) as well as a synthesis which shows how the components cooperate and the variables interact. Therefore, our approach will at first be analytic, and only later synthetic.

We begin by defining Pco_2, $[HCO_3^-]$ and pH because these are the chief acid-base variables which the body normally regulates and which in disease may require regulation by the physician. The dramatic effects of sudden changes in these variables demand that therapy be based on an appreciation of the individual performance of each—i.e. how they normally behave and how they react in acute and chronic disturbances. Indeed these three are the central characters in the theater of acid-base action, each having unique and not entirely predictable peculiarities. They are clothed in appropriately different symbols, and have entirely different units of expression, for the Pco_2 is a partial pressure in mm Hg, the $[HCO_3^-]$ is a concentration in milliequivalents per liter and the pH is a measure of the electrochemical potential of protons. And they play three entirely different roles especially with respect to their time and place of action, for the Pco_2 is regulated via the lungs, the $[HCO_3^-]$ via the kidneys and the pH via the activity of both these organs.

To emphasize the clinical and qualitative rather than the chemical and quantitative aspects of acid-base regulation, in the first three chapters we will consider these variables as the escaping tendency of carbonic acid, the bicarbonate reserve and the base:acid ratio index. The meaning of these concepts is clarified by developing the distinction between intensity and

1

quantity factors (see also Appendix II) in blood gas physiology. We will therefore study these variables as:

P_{CO_2},   an intensity factor in CO_2 transport,
$[HCO_3^-]$, a quantity factor in acid-base regulation,
pH,   an intensity factor of acidity.

The fourth variable we will deal with in detail is the arterial P_{O_2}. The study of oxygen transport by the blood beautifully illustrates the difference between an intensity factor (P_{O_2}) and quantity factors (O_2 content, % saturation). In the clinical chapters it will be evident that blood gas transport and acid-base regulation are as closely connected in disease as they are in health.

In attempting to bring these four variables together we will adopt a consistent symbolic and graphic representation. (The inherent difficulty of considering more than two variables simultaneously accounts for much of the confusion surrounding acid-base physiology; the continuing multiplicity of graphic formulations has added to it.) Even though, as we shall see, none of the four variables is completely independent of the others, the intensity factors are relatively independent of the quantity factors. Hence, intensity factors (e.g. P_{CO_2}, P_{O_2}) will nearly always be plotted on the abscissa (x-axis) of a graph, while the quantity factors (e.g. $[HCO_3^-]$, O_2 content) will be plotted on the ordinate (y-axis).

It is logical as well as traditional to place independent variables on the x-axis. The body responds to (via chemoreceptors) P_{CO_2} and P_{O_2} and rapidly manipulates these variables; the physician (via measurements made with electrodes) can respond to and rapidly manipulate these variables. Finally, as will become apparent, plotting both P_{CO_2} and P_{O_2} as independent variables on the x-axis aids in achieving an integrated approach to blood gas regulation.

1

The Escaping Tendency of Carbonic Acid, Pco₂

"In the course of a day a man of average size produces, as a result of his active metabolism, nearly two pounds of carbon dioxide. All this must be rapidly removed from the body. It is difficult to imagine by what elaborate chemical and physical devices the body could rid itself of such enormous quantities of material were it not for the fact that, in the blood, the acid can circulate partly free and, in the lungs, by a process which under ordinary circumstances has all the appearances of a simple physical phenomenon, can escape into air which is charged with but little of the gas. Were carbon dioxide not gaseous, its excretion would be the greatest of physiological tasks; were it not freely soluble, a host of the most universal existing physiological processes would be impossible." L. J. HENDERSON (1913)[2]

THE Pco₂ of the blood is a variable which can change much more rapidly than the other acid-base variables. Although expressed in mm Hg and defined as a partial pressure, it is best described as an escaping tendency—i.e. the tendency of CO_2 molecules to move from a region of high to a region of low Pco₂ as shown in Figure 1.

THE CHEMICAL NATURE OF PHYSICALLY DISSOLVED CO₂

The gas, CO_2, is easily absorbed by all body fluids. Most of it reacts with extracellular and intracellular water and is combined as bicarbonate; a small fraction remains free as dissolved or molecular CO_2. This extremely important

3

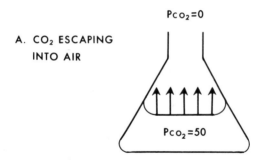

A. CO$_2$ ESCAPING
   INTO AIR

Pco_2=0

Pco_2=50

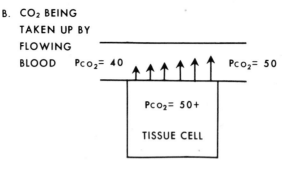

B. CO$_2$ BEING
   TAKEN UP BY
   FLOWING
   BLOOD   Pco_2= 40

Pco_2= 50

Pco_2= 50+

TISSUE CELL

FIG. 1. A. An open system in an unsteady state. B. An open system with a moving liquid phase capable of being in a steady state. The arrows represent CO$_2$ accumulation resulting from the diffusion of CO$_2$ from a region of higher to a region of lower pressure.

fraction is distributed in every body tissue almost uniformly as molecular CO$_2$ and is called Free CO$_2$.[3]

Free CO$_2$ consists almost entirely of CO$_2$ molecules which have not reacted with water and remain freely diffusible. They move rapidly through body fluids from regions of high CO$_2$ pressure to regions of low CO$_2$ pressure. In any fluid the amount of Free CO$_2$ per unit volume is directly proportional to the Pco_2 of the fluid. In a liter of venous plasma at 38°C this amount is

$$[\text{Free CO}_2] \; = \; 0.03 \; \times \quad \text{P}co_2.$$
$$\text{1.5 mEq/L} \qquad\qquad \text{50 mm Hg}$$

The number 0.03 is simply the solubility factor which relates the amount of CO$_2$ which plasma can carry as Free CO$_2$ to the CO$_2$ pressure which drives it into solution as shown in Figures 2 and 22A.

The diffusion of CO_2 across the alveolar membranes illustrates how CO_2 transport depends on the pressure to escape from the blood and how this pressure is maintained in the blood from a steady supply of Free CO_2.

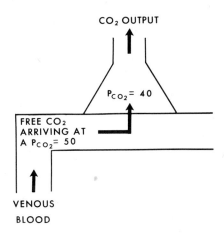

In the steady state which results when the pulmonary ventilation puts out CO_2 as fast as it is produced in the tissues, the P_{CO_2} in the alveolar gas phase of the lungs and in the blood remains almost constant.

With this qualitative preparation we will now state the quantitative definition of the P_{CO_2} in alveolar gas, often written $P_{A_{CO_2}}$:

$$P_{A_{CO_2}} = (P_B - P_{H_2O}) \times F_{A_{CO_2}}$$

where P_B is barometric pressure, P_{H_2O} is the pressure of water vapor (about 47 mm at body temperature) and $F_{A_{CO_2}}$ is the volume fraction of CO_2 in dry alveolar gas.

The P_{CO_2} of a liquid such as blood in equilibrium with a gas of given P_{CO_2} is a concept represented by the top drawing of Figure 3. However, blood P_{CO_2} although quantitatively defined by a calculation made for a gas phase with which blood is in equilibrium, is more understandable as an escaping tendency. Normally kept from varying because controlled by the ventilation, the P_{CO_2} is subject to rapid changes; this is because it measures only the free fraction, i.e. the dissolved and uncombined fraction, of the total blood carbon dioxide and this fraction can be quickly depleted in certain unsteady states. The slightest

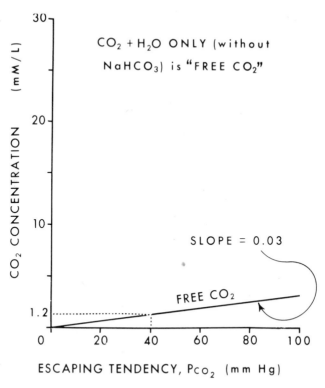

FIG. 2. [Free CO_2] is normally 1.2 mM/L in arterial plasma at 38°C since the arterial Pco_2 is normally 40 mm Hg. The linear relation between the concentration of free or molecular CO_2 and the CO_2 pressure is an example of Henry's law. The value of the slope, 0.03, is the solubility factor for CO_2 in plasma. The symbol S has been used for this factor.[4]

increase in pulmonary ventilation, if out of proportion to the rate at which CO_2 is brought to the lungs, will "wash out" or "pump out" the Free CO_2 from the blood and therefore lower the blood Pco_2 proportionally. This type of over-breathing may easily drive the Pco_2 below 35 mm Hg, producing a condition called CO_2 depletion or hypocapnia.

The concentration of Free CO_2 is denoted by placing the term in square brackets, i.e. [Free CO_2]. In the past it was often written as [H_2CO_3] and spoken of as the concentration of carbonic acid. This has the incorrect connotation that all Free CO_2 is bound to water, i.e. hydrated, whereas in fact most Free CO_2 is simply physically dissolved CO_2. Indeed only about one eight-hundredth of the Free CO_2 is in the form H_2CO_3. The quantity, [H_2CO_3] itself, cannot even be determined accurately[5] while the Pco_2 can and—in medicine—often should be.

THE PHYSIOLOGIC ROLE OF THE P_{CO_2}

The Arterial P_{CO_2}*, the Pulmonary Ventilation and the Steady State*

Because CO_2 moves so easily across lung membranes, its excretion from the body is not limited by the diffusion process of moving it from the blood to the alveolar gas. The only important limitation to CO_2 output encountered clinically is in transporting CO_2 from the lungs to the atmosphere; that is to say, CO_2 output is ventilation-dependent. The effectiveness with which pulmonary ventilation removes CO_2 from the blood is accurately reflected by the P_{CO_2} in the blood which flows from the lungs, namely, the arterial blood.

This state of affairs is often described by saying that blood "reaches equilibrium with CO_2 in alveolar gas." Now although it is true that approximate numerical equality normally exists between the P_{CO_2} in the alveolar gas in the lungs and in the blood leaving the lungs, a true equilibrium does not exist because this is a state reached only in closed or isolated systems; the system under consideration, namely, the blood while in the lungs, is an *open* system.

An open system is defined thermodynamically as one permitting the exchange of matter as well as heat with the surroundings. A closed system allows heat to be exchanged, but not matter. An isolated system is one in which neither heat nor matter can be exchanged with the surroundings. The concept of an open system is fundamental to clinical acid-base understanding.

The physiologic significance of CO_2 production in the tissues and excretion by the lungs is often neglected in textbooks on acid-base chemistry. (A notable exception is the textbook by Edsall and Wyman.[6]) Each body tissue is an open and not a closed system. Carbon dioxide is produced in every tissue cell, and diffuses through all body compartments; at least 15,000 mEq are daily delivered to and escape from the lungs of adult man. Despite this rapid flux of CO_2, the P_{CO_2} is nearly constant in any given tissue.

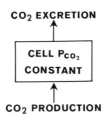

A *steady state* in this context is defined as a state in which the P_{CO_2} in any given tissue is constant; more generally a steady state is one in which variables at any point in a system do not change with time.

The Regulation of Pco₂ in Health

How is this constancy of P_{CO_2} throughout the body achieved? By feedback regulation of pulmonary ventilation, which controls arterial P_{CO_2}, and which in turn is controlled by the P_{CO_2}. The arterial P_{CO_2}-CNS-lung system is an example of a negative feedback chemostatic control mechanism: if P_{CO_2} rises, ventilation is stimulated by this rise and lowers the P_{CO_2}; if ventilation is increased for some reason, the P_{CO_2} falls and the CO_2 stimulus to breathing is reduced until the P_{CO_2} rises to its steady state level.

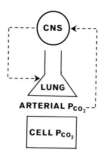

What determines the particular steady state value of P_{CO_2} found in blood? If CO_2 is only a waste product, and potentially toxic in addition, why not get rid of it as completely as possible and keep its level as low as possible? Is the "normal" value always the optimal value? These questions are important clinically and will be considered later. Here we will only lay a foundation for understanding P_{CO_2} regulation in disease.

At sea level the arterial P_{CO_2} is maintained both at rest and during moderate exercise at 40 ± 2 mm Hg—a value which is called normal. This value is not necessarily optimal under all conditions of activity or in every environment. At one mile above sea level the normal (and probably optimal) arterial P_{CO_2} is 36 mm Hg in man. The process of adaptation to high altitude which permanently lowers the arterial P_{CO_2} is dictated by other needs than CO_2 removal and involves not just the lungs but the response of the central nervous system, the bone marrow and the kidneys to low oxygen pressure. As a result of the integrative activity of these organs, men living at altitude are in a steady state of chronic respiratory ⟨alkalosis⟩, but because their blood pH is within normal sea-level limits, the word ⟨compensated⟩ is used to describe their acid-base status. This usage implies that normality of pH in the *plasma of circulating blood* is all-important. But, as we will see, the pH inside certain cells and in the cerebrospinal fluid and the state of oxygenation of brain tissue (see Chapter 6) may sometimes have to be considered even more important.

The Regulation of P_{CO_2} in Disease

Patients with chronic pulmonary disease tend to have an arterial P_{CO_2} that is higher than normal by 5 to 20 mm Hg. Their arterial blood pH is usually slightly low—between 7.30 and 7.37. Such patients have adapted to a new steady state in which the arterial P_{CO_2} is permanently high and the pH permanently low. Because of the low pH they are said to be in ⟨uncompensated respiratory acidosis⟩.

Unlike high-altitude dwellers, who may greatly change (lower) their P_{CO_2}, many patients with lung disease do not bring their arterial pH to "normal," although their renal function is adequate to bring this about.[7,8] In view of this, it may be questioned whether either arterial P_{CO_2} or pH "should" be kept normal by therapy. Optimal levels of P_{CO_2} and pH in various tissues may not be the same as normal levels in certain steady states imposed by disease.

In chronic respiratory insufficiency, for instance, neither blood pH nor the pH of cerebrospinal fluid (CSF) are "normal": as blood pH falls with increasing P_{CO_2}, the CSF pH falls "in identical proportions."[9] The finding of a chronically low CSF pH agrees with the work of some authors,[10,11] but not all.[12] In CSF studies the word ⟨compensated⟩ should be used with caution.

The state of patients requiring the greatest attention and understanding is usually not "steady" but a state in which, at least temporarily, one process is occurring faster than another. The most important example of such an unsteady state in the present discussion is the state of acute CO_2 retention. This results directly (and only) from CO_2 being produced faster than it is excreted. Eventually a new steady state is reached in which the P_{CO_2} in the blood is increased, a circumstance most often caused by partial airway obstruction as shown in Figure 3. Such obstruction leads to a reduction in ventilation which causes the rise in P_{CO_2} in the lungs and body fluids. This process is often called respiratory ⟨acidosis⟩ but the term "acute CO_2 retention" serves better for clinical purposes, being less ambiguous.

Interferences with the transport of material in the body are encountered daily in the practice of medicine. Whether the obstruction be intestinal, biliary or venous, the result will always be a build-up of *pressure* behind the obstruction, this pressure being in an unsteady state until it either stabilizes at a higher level or is relieved by attempts to overcome the obstruction. In the case of obstruction of pulmonary airways (e.g. from mucosal edema or by mucous plugging) the CO_2 retention is often aggravated in a special way. In order to overcome airway obstruction by attempts to force air through narrowed bronchi, the respiratory muscles work harder, expend more energy, and hence must use more oxygen and produce more CO_2. Clearly a point may

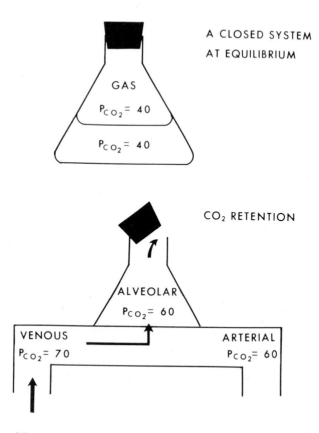

FIG. 3. When CO_2 production exceeds CO_2 excretion from a system, the P_{CO_2} rises in the system. Until the P_{CO_2} reaches and stays at a new higher level, the system is in an unsteady state.

be reached (and in very severe asthma is reached) when the effort to get rid of CO_2 by ventilating obstructed lungs is so costly that the blood P_{CO_2} rises instead of falls.

This important concept can be expressed in the approximate formula

$$\text{Arterial } P_{CO_2} \propto \frac{\text{Non-pulmonary work} + \text{Pulmonary work}}{\text{Alveolar ventilation}}$$

The formula states that for a given alveolar ventilation the arterial P_{CO_2} will rise with an increasing metabolic cost of breathing; also that for a given total metabolism, a reduced alveolar ventilation will raise the P_{CO_2}. It is difficult to exaggerate the physiologic and clinical importance of this formula, non-exact as it is.* When pulmonary work is experimentally increased, normal subjects accept increases in arterial P_{CO_2} rather than expend the effort to lower it.[13] Riley[14] provided quantitative evidence that obstructed patients, because their work of breathing is so metabolically expensive, may "choose" a lower level of ventilation than that needed to maintain blood gas levels because, as Cherniack[15] puts it, such a choice "increases the amount of oxygen available for non-ventilatory work." Thus the above simple formula helps to explain how P_{CO_2} is regulated at optimal but not necessarily normal levels in pulmonary disease, the optimal level being determined more by tissue demands for oxygen (see Chapters 4 and 6) than by the need for normality of the acid-base state of blood—a conclusion with important therapeutic implications. (When should the physician attempt to return a patient's abnormal P_{CO_2} to a normal level?)

MEASUREMENTS OF P_{CO_2} AT THE BESIDE

Arterial versus Venous Blood

Venous blood, although suitable for determining electrolyte concentrations (including total CO_2 content), is much less satisfactory than arterial blood for P_{CO_2} measurement. This is because too often a peripheral venous sample is taken while the vein is engorged with non-flowing blood. With such venous stasis, P_{CO_2} can rise 20 to 40 mm Hg in a minute or so and hence be extremely misleading. (In the case of *oxygen* measurements, venous blood is never a suitable substitute for arterial.)

Free-flowing venous or capillary blood, especially from a warmed hand or earlobe will often yield P_{CO_2} values within a few mm Hg of arterial and is satisfactory as a rough index of alveolar ventilation.

Arterial blood sampling, if done properly, is easy, safe, often of crucial importance and seldom any more painful than a venipuncture. The equipment needed is simple.

* The exact equation,

$$\text{Alveolar } P_{CO_2} = 0.863 \, \frac{CO_2 \text{ output}}{f(V_T - V_D)},$$

is difficult to apply clinically because the only variable that is easy to measure is f, the respiratory rate. Tidal volume, V_T, is being measured in respiratory care units with great advantage. The other variables, i.e. V_D, the dead space and CO_2 output, are virtually unmeasurable clinically; furthermore the concept of CO_2 output from the lungs is less important than that of CO_2 production in the tissues in unsteady states.

Both equations leave out recent evidence that a reduced ("blunted") hypoxic ventilatory drive can contribute to chronic hypoventilation and that hypercapnia, of itself, may reduce oxygen requirements and CO_2 production under certain conditions (see Chapter 6).

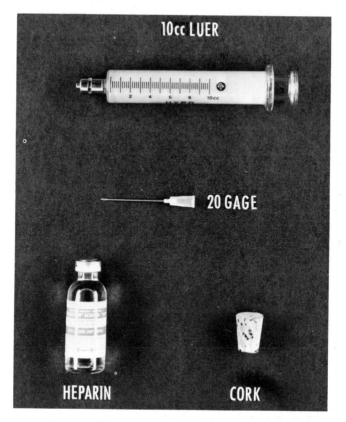

If a Pco_2 measurement is needed, both pH and Po_2 determinations are likely to be even more urgently needed. Hence arterial blood which is suitable for all three of these direct measurements is always to be preferred.

TABLE 1. *Normal values of blood Pco_2 in mm Hg**	
Arterial	*Venous (peripheral)*
40 ± 2	46 ± 4

* Mean and approximate S.D. values given; means about 4 mm Hg lower at one mile above sea level.

The sicker the patient, the less suitable the peripheral venous sample becomes since peripheral vein blood becomes less and less representative of arterial blood as the circulation fails. In shocked patients, central venous blood analyses may be useful if properly interpreted,[16] but the most generally useful sites for obtaining samples for blood gas and acid-base analyses are the brachial and radial arteries—in children as well as in adults.[17,18]

Techniques for Measuring P_{CO_2} (All Dependent on pH)

Because blood P_{CO_2} is a partial pressure, it cannot be measured by knowing only the total CO_2 content of a blood or serum sample. It can, however, be accurately calculated via the Henderson-Hasselbalch equation (see Chapter 3 for its derivation) from the CO_2 content of plasma if the pH of the sample is also known.

Why, then, have the last several generations of physicians been unable to obtain an accurate blood P_{CO_2} on their patients? Simply because it is only recently that clinical pH meters have become both accurate and precise, for the pH meter is the essential component of the three most generally useful methods of determining blood P_{CO_2}. These meters can be operated successfully at the bedside or in the next room—an important advantage since the pH of blood, unless the syringe containing the blood is immediately put in an ice bath, falls significantly in a few minutes.

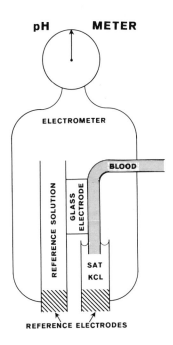

1. *Calculation of* P_{CO_2} *from serum or plasma* CO_2 *content and blood pH.* This method is the time-honored one of calculation by the Henderson-Hasselbalch equation or by the use of nomograms. The CO_2 content of an anaerobically collected blood sample is determined by the Van Slyke gas extraction technique,

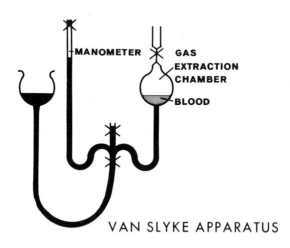

VAN SLYKE APPARATUS

by the Natelson technique, or by any of the automated methods available in clinical laboratories. The pH is determined in a blood sample collected anaerobically in a heparinized syringe or other suitable vessel containing heparin (not citrate or oxalate). This pH, though often called "blood pH" is really the pH of true plasma (i.e. plasma in contact with red cells) and can therefore be used in the following equations for calculating the blood P_{CO_2} (in mm Hg) from the [Total CO_2] in plasma or serum (in mM/L) when both the patient and the pH electrode are at 38°C:

$$pH = 6.1 + \log \frac{[\text{Total } CO_2] - .03\ P_{CO_2}}{0.03\ P_{CO_2}}$$

or

$$P_{CO_2} = \frac{[\text{Total } CO_2]}{0.03\,[10^{pH-6.1} + 1]}.$$

These equations are the basis of many time-saving nomograms,[4,19,20] the use of which is always better understood if the nomographic answers are compared with those obtained by pencil and paper, due consideration being given to temperature effects and to the (generally slight) variations from the usually given value of 6.1 for the pK of the P_{CO_2}-bicarbonate system in normal human plasma.[4,21]

2. *The Astrup tonometer technique for obtaining* P_{CO_2}. This is a simple,

reliable technique for obtaining the P_{CO_2} of blood. It is not the same as the Astrup "New Approach."[22] The technique does not and need not involve the "Base Excess" concept in any way. (See Appendix III.)

A blood specimen is divided into three portions; two are shaken in Astrup "tonometers," i.e. vessels in which P_{CO_2} can be controlled at known values. The three portions of blood are then analyzed for pH:

at a known P_{CO_2} probably higher than the unknown P_{CO_2};
at a known P_{CO_2} probably lower than the unknown P_{CO_2};
at the unknown P_{CO_2} of the anaerobically obtained sample.

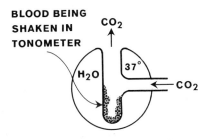

BLOOD BEING SHAKEN IN TONOMETER

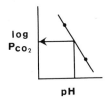

SIGGAARD–ANDERSEN NOMOGRAM

Since an almost exactly linear inverse relationship exists between pH and log P_{CO_2} of true plasma, it is easy to find the unknown P_{CO_2} by graphic means. An excellent description of the technique has been recently published.[23] A part-time technician can be expected to produce reliable results; if properly trained, a Pulmonary Fellow or Resident can do likewise.

3. *The Severinghaus P_{CO_2} electrode.* This consists of a plastic membrane separating a blood sample from a bicarbonate solution containing a pH electrode. CO_2 from the sample (but not HCO_3^-) diffuses through the membrane. At equilibrium (reached in a minute or so, depending on the geometry of the system), the log P_{CO_2} is an inverse linear function of the pH of the bicarbonate solution by the Henderson-Hasselbalch equation. The device responds to pH, but the scale on the meter is calibrated to indicate the P_{CO_2} directly.

SEVERINGHAUS

CO$_2$ ELECTRODE

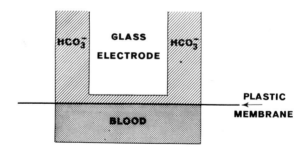

This method has been described by Severinghaus,[4] and by Winters *et al.*[23] It requires a full-time technician to guarantee reliable results because of the care with which the equipment—especially the membrane—must be maintained. Physicians and part-time technicians generally will have trouble with this technique. (It is not, of course, the *time* spent in the blood gas laboratory but the care with which the equipment is maintained which is important.)

THE CLINICAL SIGNIFICANCE OF P$_{CO_2}$

Why measure P$_{CO_2}$ except in patients in whom pulmonary disease is suspected? What, in a community hospital say, is the advantage of measuring anything more than a CO_2 content or CO_2-combining power? A partial answer is "Very little unless other indices of acid-base disturbances are understood." Since these other indices (pH, [HCO_3^-], P$_{O_2}$) can only be understood by being considered together, the full answer to these questions is reserved for the chapters to come. Here we will focus on the clinical value of the arterial P$_{CO_2}$ considered as an isolated measurement.

An Increased Arterial P$_{CO_2}$ Is a Quantitative Danger Signal

Since the lack of oxygen is almost immediately lethal, it is not surprising that the body vigilantly guards against threats to its oxygen supply. The very location of the carotid and aortic bodies, which respond to arterial hypoxemia, suggests that they act as sentinels protecting the vital centers from oxygen lack. What is surprising—at first glance—is that chemoreceptive sentinels are far more responsive to increases in arterial P$_{CO_2}$ than to hypoxemia.

The first glance at a biologic system is always surprising to a pure mechanist. It is clear from the point of view of evolutionary respiratory engineering that, in air-breathing animals, the oxygen supply to the brain is better safeguarded by an alarm system triggered by the ubiquitous intensity factor of CO_2 transport, i.e. the "smoke" of metabolism, than by a mechanism which only goes off when the "fire" is about to go out from lack of oxygen. (See also Haldane,[1] Chap. XIV.) For example, inhaling CO_2-enriched air causes immediate hyperventilation, as a response to the increase in blood Pco_2, or as we shall see later, a fall in pH in the blood perfusing the medullary and peripheral receptors or, more likely, the pH near or in the cells of these receptors. The respiration doubles when arterial Pco_2 is increased by only a few mm Hg. However, man must breathe air very low in oxygen before a similar augmentation in ventilation occurs: "most chemoreceptors . . . are not stimulated effectively or maximally until the [arterial] Po_2 has fallen to 40 or 50 mm Hg (equivalent to the inhalation of 8 to 10% oxygen)."[24] The last two statements oversimplify the situation since the CO_2 and O_2 stimuli actually potentiate each other as shown in Chapter 6.

It is clinically important to know that:

1. The physiologic danger signal which usually first warns of ventilatory failure is an increase in the blood Pco_2, that
2. The sensitivity to this signal depends on the undrugged activity of the nervous system, that
3. Only in dire circumstances does the body depend on hypoxemia itself as the stimulus to breathe, and that
4. A Pco_2 alteration of days' to weeks' duration has an entirely different significance from an acute Pco_2 change.

These four physiologic facts lead in turn to four clinical conclusions:

1. If the physician is to approach the effectiveness with which the patient's chemoreceptors recognize ventilatory failure, he must *measure* the blood Pco_2 and not depend on the vague and unreliable clinical signs of CO_2 retention.
2. If the physician is to preserve the patient's defenses against respiratory failure, he must be careful in administering sedatives which depress the normal sensitivity to high Pco_2 and force the organism to depend on the less sensitive reactions to low Po_2.
3. If the physician waits for the patient to become blue to diagnose ventilatory failure, he has waited too long.
4. The best way to decide if a Pco_2 change is acute or chronic (short of prolonged observation and multiple determinations) is to measure other acid-base variables, e.g. $[HCO_3^-]$ or pH.

We are thus forced to end this chapter's attempt to view Pco_2 as an isolated measurement and to begin to place it in dynamic relation to other measurements.

2

The Bicarbonate
Reserve, [HCO₃⁻]

"Bicarbonates constitute the first reserve, so to speak, in neutralizing acid. They are effective in a far greater degree than the salts of any acids in equal concentration could be because of the regulation of carbonic acid concentration by diffusion and excretion. . . . Both intracellular and extracellular aqueous solutions of the organism . . . acting selectively as reservoirs of supply and as vehicles of escape . . . surpass the efficiency of any possible closed aqueous solutions of like concentration in preserving hydrogen ion concentration."

L. J. HENDERSON (1908)[25]

THE BICARBONATE ion, HCO_3^-, is the chemical species which comprises almost all the CO_2 produced in the body. It can be thought of as the quantity factor of CO_2 transport as contrasted to the intensity factor, P_{CO_2}. The serum bicarbonate normally contains nineteen twentieths of the total CO_2 concentration and is the combined form of serum CO_2 in contrast to Free or molecular CO_2. The total CO_2 content is the sum of these two species. In normal plasma or serum

$$\underset{25.2}{[\text{Total } CO_2]} = \underset{1.2}{\overset{\text{Free}}{[0.03 \; P_{CO_2}]}} + \underset{24}{\overset{\text{Combined}}{[HCO_3^-]}} \qquad \text{mEq/L}$$

where the numbers indicate normal arterial concentrations.

The symbol $[HCO_3^-]$ here denotes the number of milliequivalents of bicarbonate ions contained in a liter of plasma. Of late it is often referred to as the "actual bicarbonate" to distinguish it from various chemically manipulated samples, the oldest of which is the "CO_2-combining power." Another example is called the "standard bicarbonate." These terms are not used in the present text but are dealt with in Appendix III.

Carbon dioxide is also combined in small amounts as $CO_3^=$ and as carbamino compounds, but these can be neglected in considering clinical acid-base disturbances.

Bicarbonate concentration will be expressed in mEq/L. Although the "equivalent" as a unit is "ambiguous and unnecessary,"[26] it is so well established that we will use it until the mole is more generally accepted as the *"quantity name"* for an amount of substance.

With certain exceptions, a lowering of plasma $[HCO_3^-]$ indicates that bicarbonate has been partly used up and hence partly destroyed by fixed acid invading the blood stream. This chapter will explain both this process of destruction and the exceptional cases in which $[HCO_3^-]$ changes do not reflect fixed acid changes.

THE CHEMICAL NATURE OF THE BICARBONATE ION

In contrast to the simplicity of Free CO_2, the HCO_3^- ion has a complex character and plays two major chemical roles.

1. As part of the CO_2 transport system, bicarbonate is generated in the tissue capillaries and some of it is broken down or destroyed in the lungs. The generation and destruction of HCO_3^- always involves *H^+-transfer*, the central process in every acid-base reaction.[5] The behavior of HCO_3^- led Pitts[27] to write: "Bicarbonate is a unique ion, in that it has no permanence. Its existence is fleeting . . . a step in the transfer of carbon dioxide between cells and lungs."

2. Bicarbonate is part of a marvelously efficient defense system against pH change. The plasma $[HCO_3^-]$ is what Van Slyke "called the *alkali reserve*, an appellation which it deserves because it is a uniquely available form of alkali and also because it mirrors more or less closely, as will be shown, the reserve of available alkali present in the body as a whole."[28]

We will show that the key to understanding how excess ionizable hydrogen is disposed of by the body requires a full realization of how the central process of H^+-transfer *generates bicarbonate*. We define two important words as follows:

Transfer—a chemical process involving a redistribution of bound material such as potentially ionizable hydrogen among available binding sites.

Transport—a physical or biologic process involving the movement of material through a solution or through the body.

The Bicarbonate Ion as a CO_2 Transport Vehicle

If CO_2 is produced in a cell, it reacts with water (rapidly or slowly, depending on whether the enzyme carbonic anhydrase [c.a.] is present or absent) as follows:

$$CO_2 + H_2O \xrightarrow{\text{c.a.}} H_2CO_3 \rightarrow H^+ + HCO_3^-$$
$$\text{Free } CO_2 \qquad\qquad\qquad \text{Combined}$$

As we mentioned in the first chapter, the concentration of H_2CO_3 is only about one eight-hundredth of that of free molecular CO_2; we will therefore omit it from most of our formulations. The reaction is reversible and is often written with double arrows to signify the state of equilibrium.

A qualitative notion of the reaction between CO_2 and *pure water* when the Pco_2 is held at 40 mm Hg is as follows where the reaction is said to be "to the left":

add

$Pco_2 = 40$

$CO_2 + H_2O \underset{\longleftarrow}{\overset{(H_2CO_3)}{\longrightarrow}} H^+ + HCO_3^-$

pH = 4.6

What does this mean? The heavy bars indicate that H^+ and HCO_3^- are "blocked," having "no place to go." Note that an important consequence of this is the piling up of H^+ causing pH to fall to 4.6. Another important consequence is that water alone cannot carry much CO_2. Thus at a Pco_2 of 40 mm Hg, pure water at 40°C dissolves only 1.225 mEq/L of CO_2 in all forms (1.2 free and only 0.025 as HCO_3^-).*

How is it then that [HCO_3^-] reaches such high levels in the body fluids when CO_2 in pure water yields scarcely any HCO_3^- ions? This is because H^+ is

---

* The traditional way of indicating these relationships quantitatively is: "The K of carbonic acid, which is relatively weak, is only about 5×10^{-7} at body temperature. In consequence less than 0.2% of the carbonic acid . . . is dissociated into H^+ and HCO_3^- ions."[28] To check the figures, converting to molar concentrations, we see that $K = [H^+][HCO_3^-]/[\text{Free } CO_2] = (2.5 \times 10^{-5})^2/0.0012 = 5.2 \times 10^{-7}$, and $-\log 2.5 \times 10^{-5} = 4.6$.

The temperature of 40°C is chosen in this example to bring the solubility of CO_2 in pure water closer to its solubility in plasma at 38°C so that in both systems [Free CO_2] in mEq/L = 0.03 Pco_2.

provided by body fluid ⟨bases⟩ with "a place to go." This means that body fluid bases accept (bind) H^+ ions, allowing the above reaction to go to the right, thus generating bicarbonate. In brief, H^+ is *transferred* from the CO_2-bicarbonate system to another system.

To understand this we will proceed in logical steps to build what may be called the acid-base "structure" of plasma.

First we examine what happens when NaOH is in a water solution in the presence of CO_2 gas. This process can be visualized on the left of Figure 4. Here it is clear that CO_2 is absorbed by Na^+OH^- solution. However, the fundamental process of H^+ transfer to the ⟨base⟩ (i.e. OH^-) is not often written in such explicit notation as that in the middle of Figure 4. The traditional way of describing the chemical reaction is

$$NaOH + H_2CO_3 \rightleftarrows NaHCO_3 + H_2O,$$

i.e. in a notation which "hides" two important processes, namely, proton transfer, a chemical process, and CO_2 transport, a physical process. The reaction is better depicted for our purposes by showing explicitly that H^+ (hydrogen nuclei or protons) are transferred between systems, i.e.

A SOURCE OF CO₂ GAS

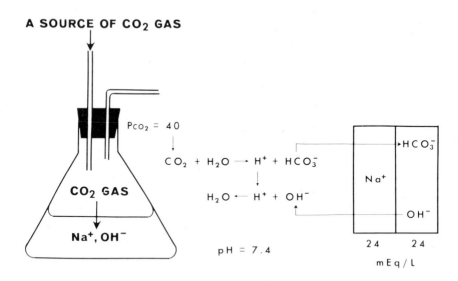

FIG. 4. When CO_2 reacts with sodium hydroxide, HCO_3^- is generated and OH^- is used up. On the far right, 24 mEq/L of Na^+ are matched against a progressively increasing concentration of HCO_3^- and progressively decreasing concentration of OH^-. After equilibrium has been reached, the concentration of H^+ is very low and pH = 7.4.

Sodium hydroxide is written Na^+,OH^- to indicate that in water its ions are *matched* not combined.

A source of CO_2 gas

$\downarrow$

$CO_2 + H_2O \rightarrow H^+ + HCO_3^-$ (is generated).

$\downarrow$

Na^+,OH^-

The addition of OH^-, which pulls the reaction to the right, *generates* HCO_3^-. This is an extremely important process. The CO_2-Na^+,OH^- reaction is very similar to the process of ⟨buffering⟩ which occurs in the blood except that it cannot be easily reversed. The OH^- has bound the H^+ too tightly for this, that is, Na^+,OH^- is too "strong" a base.

The near irreversibility (using this word loosely) of the reaction between CO_2 and Na^+,OH^- makes a sodium bicarbonate solution a poor vehicle for CO_2 transport as compared with blood. This can be illustrated by CO_2 uptake or CO_2 dissociation curves.

We have met one such CO_2-uptake curve in Figure 2 which showed how small a quantity of CO_2 is absorbed by pure water. Figure 5 illustrates how much more CO_2 is taken up by Na^+,OH^- solution than by water. It can be said that the alkali provides Na^+ for HCO_3^- to be matched against; it is much more important to say that OH^- was provided to bind H^+.

The height of the HCO_3^- curve is a measure of the fact that about twenty times more CO_2 is held as HCO_3^- than is present as Free CO_2. However, it can be seen that as P_{CO_2} is increased above 20 mm Hg, no more HCO_3^- is added because its curve is virtually horizontal. The only measurable CO_2 taken up as a result of further P_{CO_2} increase is in the form of Free CO_2. Thus we see that although this solution can *hold* the same amount of CO_2 as is normally present in plasma, it cannot *vary* this amount significantly. In other words, the height of the HCO_3^- curve reflects a considerable ability to take up and carry CO_2 at a given P_{CO_2}, but its horizontal nature indicates that the solution would be almost useless as a CO_2 transport system because it reflects a nearly irreversible process.

In blood, however, the process is easily reversible. This is because H^+ is reversibly bound—not by the strong base Na^+,OH^- but by hemoglobin. This makes possible HCO_3^- destruction as well as HCO_3^- generation.

While blood is in the lungs the reaction between CO_2 and H_2O is driven (pulled) to the left: as the lungs excrete CO_2, H^+ is transferred from hemoglobin:

lungs

$\uparrow$

$CO_2 + H_2O \leftarrow H^+ + HCO_3^-$.

$\uparrow$

H^+ (from hemoglobin)

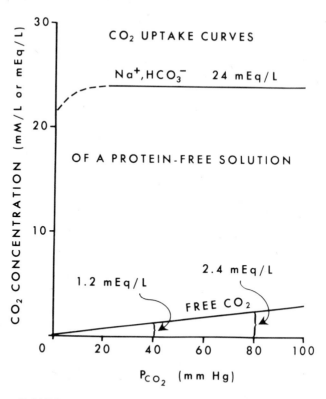

FIG. 5. A pure Na^+, HCO_3^- solution is inefficient as a CO_2 transport system. For example, an increase of Pco_2 from 40 to 80 only adds 1.2 mEq/L of Free CO_2. A negligible quantity of CO_2 is taken up as HCO_3^-, i.e. that quantity by which $[H^+]$ increases; even though $[H^+]$ doubles, the extra CO_2 taken up as HCO_3^- is still a negligible quantity. In this example:

At $Pco_2 = 40$,

$$\begin{array}{cccc} & Na^+ & + HCO_3^- & 24\text{ mEq/L} \\ & & \Updownarrow & \\ CO_2 + H_2O \rightleftarrows & H^+ & + HCO_3^- & \\ 1.2 & 0.000025 & 0.000025 & \text{mEq/L;} \end{array}$$

At $Pco_2 = 80$,

$$\begin{array}{cccc} & Na^+ & + HCO_3^- & 24\text{ mEq/L} \\ & & \Updownarrow & \\ CO_2 + H_2O \rightleftarrows & H^+ & + HCO_3^- & \\ 2.4 & 0.000050 & 0.000050 & \text{mEq/L.} \end{array}$$

Conversely, in the body tissues, the reaction is driven to the right and protons are transferred to hemoglobin:

$$\overset{\text{tissues}}{\underset{\downarrow}{}}$$

$$CO_2 + H_2O \rightarrow H^+ + HCO_3^- \text{ (increases in the blood)}$$
$$\downarrow$$
$$HHb \leftarrow H^+ + Hb^-.$$

The process which makes CO_2 transport efficient is the same process by which ⟨buffering⟩ occurs, namely, the *transfer of H^+ from one system to another.* Because of the importance of this concept—the basis of the Bronsted-Lowry theory—it deserves to be emphasized in a special way. In this book, the movement of material between systems is symbolized by a vertical arrow.

The vertical arrow does not indicate a chemical reaction nor the formation of a new chemical entity. It indicates how a chemical species in one system affects another system (in which this same species is present), either by mass action, or by chemical *transfer* or by physical *transport* of this species, e.g. H^+.

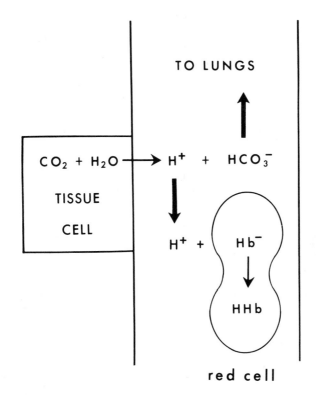

FIG. 6. In body tissues, CO_2 transport from tissue cell to blood in the capillary is facilitated by hemoglobin which removes H^+ from the Pco_2—bicarbonate system. (For clarity, many details have been left out. For instance, CO_2 actually enters red cells where carbonic anhydrase speeds its hydration and hence HCO_3^- generation; the HCO_3^- generated in red cells migrates to the plasma in exchange for Cl^-—the "chloride shift." This beautiful but intricate process has been described in detail by Roughton.[29]) The main result is that a large quantity of H^+ is transferred even though $[H^+]$ scarcely changes; hence a large quantity of CO_2 is transported with little change in Pco_2.

The most important idea in this book is implied by the oppositely directed arrows, namely, that of *vectorially directed transport.* See pp. 23, 24, 27, 28, 114 and 117 and Figures 9, 10, 13, 16, 47 and 49.

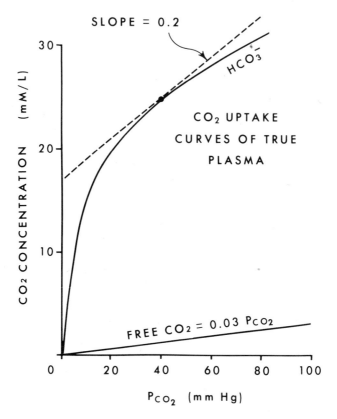

FIG. 7. The "true" plasma of whole blood, because it is in contact with red blood cells, has a steep uptake curve because hemoglobin mops up H^+ ions as Pco_2 is increased allowing HCO_3^- to be generated (and because hemoglobin donates H^+ ions as Pco_2 is reduced allowing HCO_3^- to be destroyed). The curve is plotted (and extrapolated) from the data of Table III, p. 388, in Henderson, *et al.*[30] obtained from tonometric measurements of the blood of A. V. Bock. When Bock's plasma $[HCO_3^-]$ is calculated from the Henderson-Hasselbalch equation and plotted against pH, a curve slightly convex to the pH axis is obtained. The "buffer slope" or $d[HCO_3^-]/dpH$ between pH 7.3 and 7.5 is 29 slykes. (The normal value for the buffer slope in slykes has been discussed elsewhere.[63,65])

In venous blood a large increase in plasma $[HCO_3^-]$ is possible because a large number of protons are mopped up by hemoglobin (Fig. 6). The exact relationship between these quantities will be taken up later—after the Henderson-Hasselbalch equation has been derived. For the present it is important to focus on the fact that the $[HCO_3^-]$ change depends on the quantity of H^+ transferred to and from hemoglobin and not simply on a change in $[H^+]$.

Precisely the same concepts are reflected by the steepness of the CO_2 uptake curve of blood plasma as contrasted to the horizontal nature of the protein-free bicarbonate curve. Comparison of Figure 7 with Figure 5 shows the remarkable difference between normal plasma and an aqueous bicarbonate solution. The CO_2 uptake curve for plasma, because of the effect of red cells, in made steep; in the physiologic range, a change of about 5 mm Hg in P_{CO_2} changes the bicarbonate by one milliequivalent.

In the tissues then, where CO_2 is produced, the most important chemical mechanism for carrying away this waste product is that of increasing the CO_2 absorbing power of plasma by transferring H^+ from the P_{CO_2}-bicarbonate system to the hemoglobin system. This mechanism simultaneously allows $[HCO_3^-]$ to increase in the blood with increasing P_{CO_2} and minimizes the increase of blood $[H^+]$. The ability of blood to take up CO_2 by generating bicarbonate is thus facilitated by the transfer of protons in a buffering reaction. (See also Fig. 9.)

Bicarbonate as an H⁺-Binder

The role of bicarbonate of greatest interest to physicians is that of a ⟨base⟩ or H⁺-binder. Generated continually from CO_2 and by the action of hemoglobin, it becomes, both because of its high concentration (25 mEq/L in normal venous plasma) and for a special reason we will now demonstrate, the most important base in clinical medicine.

The reaction symbolized above can now be written to show that amost all HCO_3^- is (matched against Na^+) in a reservoir, as it were. Scarcely any is matched against H^+.

$$Na^+ + \boxed{HCO_3^-} \quad \text{Venous reservoir 25 mEq/L}$$
$$\text{Arterial reservoir 24 mEq/L}$$
$$\uparrow$$
$$CO_2 + H_2O \rightarrow H^+ + HCO_3^- \quad \text{Less than 0.00004 mEq/L}$$
$$\downarrow$$
$$\text{to hemoglobin}$$

As the blood flows through the tissues, the plasma $[HCO_3^-]$ reservoir increases from 24 to 25 mEq/L. For every mEq of $[HCO_3^-]$ so generated, one mEq of H^+ ions must be transferred to hemoglobin. This formulation illustrates the essential fact that, although H^+ concentration is at all times so low that "the hydrogen ion itself has little actual physical significance,"[31] acid-base systems can transfer a large quantity of hydrogen between them. It also clearly suggests how, by simply reversing the process for transferring H^+ ions produced normally, the P_{CO_2}-bicarbonate system can act as an immediate defense against abnormal added loads of H^+ ions (as from β-hydroxybutyric acid in diabetes): HCO_3^-, the bicarbonate reserve, simply binds H^+ ions as they are formed.

The special reason, referred to above, why HCO_3^- is the most important H^+-binder in the body, is that it is part of the open Pco_2-bicarbonate system, i.e. a system from which carbonic acid, a volatile acid, can escape even with no change in Pco_2. This explains what Van Slyke meant by the "availability"[28] of HCO_3^-, i.e. the fact that initially all of it can be used up (destroyed, sacrificed) in binding H^+ ions. A qualitative (and intuitively evident) formulation of this follows:

$$\text{(Constant)} \boxed{Pco_2} \quad \text{(Reservoir)} \boxed{HCO_3^-} \Big\} \text{ Open}$$
$$\uparrow \qquad\qquad \downarrow \quad\quad \text{System}$$
$$CO_2 + H_2O \leftarrow H^+ + HCO_3^- \Big\}$$
$$\uparrow$$

$$\text{ADDED } \beta\text{-HYDROXYBUTYRIC ACID}$$
$$\downarrow$$

$$\boxed{\begin{array}{c} H\,Buf \leftarrow H^+ + Buf^- \\ \uparrow \\ Buf^- \end{array}} \Big\} \begin{array}{l} \text{Closed} \\ \text{System} \end{array}$$

It is clear that most of an organic acid load will be buffered by bicarbonate because its reaction with H^+ is free to go to the left (in fact is *pushed* or *forced* to the left by the added acid) so long as any HCO_3^- remains in the reservoir to be used up; but the non-bicarbonate buffers are part of a closed system (as if in a closed box) and therefore cannot be so used up. In other words most of the added H^+ will be bound by HCO_3^- in the open system and end up in H_2O rather than reacting with Buf^- (mostly hemoglobin) in the closed system.

All non-bicarbonate buffer bases in the closed system (e.g. Hb^-, $HPO_4^=$, etc.) are lumped together as Buf^- after the manner of Singer and Hastings,[32] and Winters et al.[23] Strictly speaking the Buf system is not completely closed since H^+ can enter and leave it; the essential point is that the sum [H Buf] + [Buf$^-$] remains constant in whole blood in contrast to the [Total CO_2] which varies with pH. In other words the material in the closed system is fixed in total quantity; the material in the open system can be lowered in total quantity when "pushed" to the left by fixed acid.

The above formulation also shows how *pulling* the open system to the left by enlistment of the nervous system (via pH-sensitive chemoreceptors) might greatly increase this buffering. If the Pco_2 can be lowered—and it is lowered by the hyperventilation regularly accompanying organic acid invasion —an even greater proportion of the added H^+ will become tied up in H_2O rather than in H Buf. The lowering of Pco_2 by CNS and pulmonary enlistment is often called "respiratory compensation."

In addition to the essentially chemical process of H^+ transfer between chemical systems, potentially ionizable hydrogen, symbolized by H, is physically trans*ported* from one body compartment to another. Even at the cellular level the process of moving this hydrogen through the body fluids always involves both HCO_3^- and the properties of the open P_{CO_2}-bicarbonate system. We will therefore consider the bicarbonate reserve in the larger context indicated at the beginning of this chapter, and illustrate, using current knowledge of body fluid compartments and renal function, what Henderson[25] meant by considering the body fluid compartments "as reservoirs of supply and as vehicles of escape."

FACTORS AFFECTING [HCO_3^-] AND ITS REGULATION

In contrast to P_{CO_2} which represents easily diffusing CO_2 molecules and to which chemoreceptors are acutely sensitive, the bicarbonate ion is a bulky negatively charged ion which moves with difficulty through tissue. It is known, however, to migrate passively across the red cell membrane (in exchange for chloride in the "chloride shift") and to move between plasma and extravascular fluid, a translocation of some clinical and great theoretical importance.[33] There is some evidence that bicarbonate can be actively transported through biologic membranes,[12,34,35] but, in general, it is more a controlled than a controlling variable in the acid-base regulating system.

The Distribution of the Bicarbonate Ion in the Body

Since bicarbonate is generated from CO_2 and this gas is produced in the cells of the body by the reactions

$$CO_2 + H_2O \xrightarrow[\text{hydration}]{\substack{\text{carbonic anhydrase} \\ \text{facilitated}}} H_2CO_3 \xrightarrow[\text{dissociation}]{\substack{\text{inherently} \\ \text{fast}}} H^+ + HCO_3^-,$$

it is clear that HCO_3^- is present in every tissue. Its concentration depends largely on the anatomic sites of location of the enzyme *carbonic anhydrase.* This enzyme, as we have noted, is present in red blood cells where it speeds HCO_3^- generation in the systemic capillaries and destruction in the pulmonary capillaries; its action in the kidney, as we shall see, is important in regulating the [HCO_3^-] reserve of the body.

In body fluids such as plasma, and extravascular fluids such as lymph and cerebrospinal fluid, the [HCO_3^-] can be accurately computed because in these homogeneous substances pH and P_{CO_2} can both be accurately measured. The Henderson-Hasselbalch equation

$$pH = pK + \log \frac{[HCO_3^-]}{S \cdot P_{CO_2}}$$

TABLE 2.   *Normal [HCO_3^-] values (mEq/L)*

Arterial Plasma	Venous Plasma	Interstitial Fluid	Cerebrospinal Fluid
24	25	29	23

(The standard deviations of most reported series of normal values are about ±2 mEq/L.)

is used to compute [HCO_3^-] in these fluids, allowances being made for variations in pK and for different solubilities, S, of CO_2 in different fluids. (In interstitial fluid, for example, the bicarbonate concentration is somewhat higher than in plasma partly because of the Gibbs-Donnan effect; in CSF it is lower for more complex reasons.) (See Table 2.)

For the inside of intact cells the only term in this equation which is known is the Pco_2 and therefore intracellular bicarbonate cannot be calculated accurately. From estimates of the other terms, the [HCO_3^-] in the water of body cells is in the range of 12 to 16 mEq/L[36,37] on the average, about half the concentration in plasma and extravascular interstitial fluid. The total available HCO_3^- of the body is about 1000 mEq. Several times this amount is sequestered mostly as carbonate in the bones from where it can be released in response to chronically increased blood acidity.

The *immediately* available bicarbonate reserve consists of about 450 mEq of HCO_3^- in 15 liters of extracellular fluid. This bicarbonate is able to move (probably passively, by diffusion) between the 3 liters of plasma and the 12 liters of interstitial fluid. The plasma [HCO_3^-] is widely used as an index of ⟨metabolic⟩ acid-base disturbances, i.e. accumulation or loss of fixed (non-carbonic) acid from the body as a whole. To realize how well this index reflects fixed acid changes requires an understanding of how distributional factors affect plasma [HCO_3^-].

Plasma [HCO_3^-] as a Measure of Acid Accumulation in the Body

1. *Effect of fixed (non-carbonic) acid accumulation on plasma [HCO_3^-].*   When an acid such as HCl is infused into the veins of an intact animal, some of the added H^+ ions react with and hence use up some of the HCO_3^- in the blood, but many H^+ ions leave the blood stream and enter the extravascular spaces and the cells as well (Fig. 8). It is known experimentally[27,38,39] that only about one sixth of added strong acid is buffered in the blood itself, and that fully

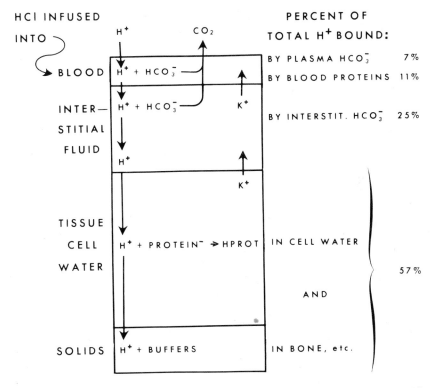

FIG. 8. When a fixed acid is added to the blood, only about 7% of the H⁺ ions are bound by plasma HCO_3^-. More than half are buffered in tissue cells. However, the quantity of HCO_3^- used up in the blood is approximately proportional to the depletion of total body buffer stores. Note how K⁺ tends to move oppositely to H⁺.

25% of the added H⁺ is bound by interstitial bicarbonate, the balance entering the cells and reacting with cellular buffers, probably mostly proteins.

Does this mean that a decrease in plasma [HCO₃⁻] is untrustworthy as an index of the severity of a fixed acid disturbance? It is true that the number of mEq by which plasma [HCO₃⁻] decreases must be less than the number of mEq of such acid added to the body as a whole: obviously extra- and probably intracellular HCO₃⁻ participates in the H⁺-binding process. Fortunately, however, the *percentage* reduction in plasma [HCO₃⁻] approximately indicates the *percentage* reduction in total body buffer stores. This follows from the work of Schwartz *et al.*,[39] who showed that "the distribution of [added] hydrogen ions among the body buffers is not affected by the magnitude of the acid load. . . ." This is to say that, if fixed acid is added, about the same proportion of the total is always buffered by the plasma bicarbonate, regardless of how much acid is added. (It is not the same as saying that, if the plasma $[HCO_3^-]$ is low, a fixed acid has been added. As we shall see, $[HCO_3^-]$ can be lowered by other means than by reacting it with fixed acid.)

In summary, a decrease in plasma $[HCO_3^-]$ often indicates that fixed acid has been added to the body in an amount reflected by the per cent decrease in $[HCO_3^-]$.

2. Effect of carbonic acid accumulation (CO_2 retention) on plasma $[HCO_3^-]$.
The fact that HCO_3^- is generated from CO_2 does not mean that the bicarbonate level is *determined* by P_{CO_2}. The expression

$$CO_2 + H_2O \rightarrow H_2CO_3 \rightarrow H^+ + HCO_3^-$$

is misleading in that without another H^+-accepting (basic) system, HCO_3^- is, in fact, generated in very small quantity, as we have seen, from CO_2 and pure water (p. 21).

It is sometimes believed that HCO_3^- is determined primarily and directly by "respiratory changes," that is, changes in P_{CO_2}. This is not true. The effect of P_{CO_2} changes *per se* in generating bicarbonate is relatively small. Attempts to "correct" a clinically measured change in plasma $[HCO_3^-]$ for the effect of a P_{CO_2} change (starting with the first attempt by Hasselbalch[40]) have led to one of the biggest bugaboos in understanding clinical acid-base physiology, namely, the notion that $[HCO_3^-]$ cannot be understood as a measured variable and that derived indices like "base excess" are needed to clarify the subject. (See Appendix III.)

The change in $[HCO_3^-]$ in a sample of plasma caused by a P_{CO_2} change is quantitatively given by the CO_2 dissociation curve appropriate to the sample in question. As we have shown (Fig. 7), at a P_{CO_2} of 40 the slope of the classical CO_2 dissociation curve in the physiologic range is about 0.2 mEq/L $[HCO_3^-]$ for every mm Hg P_{CO_2} change. In other words it takes a P_{CO_2} change of 5 mm to change $[HCO_3^-]$ one mEq/L, i.e. quite a large P_{CO_2} change (originating in the open P_{CO_2}-bicarbonate system) to change $[HCO_3^-]$ very much:

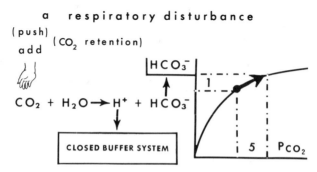

This is because the open system is *pushed* to the right (not pulled). Hence the HCO_3^- generation is limited by the capacity of the closed system to accept

H^+. In other words, the reaction can only move along the CO_2 dissociation curve. This is called a respiratory disturbance—in this case, a process which can be called respiratory ⟨acidosis⟩ but is better called CO_2 retention.

Plasma [HCO_3^-] as a Measure of Fixed Acid Loss

In contrast to the effect of Pco_2 changes, fixed acid changes greatly affect [HCO_3^-]. In vomiting (from pyloric obstruction for example), HCl may be lost from the body. In this case, [HCO_3^-] in the plasma rises precipitously.

a   non—respiratory   disturbance

$$CO_2 + H_2O \longrightarrow H^+ + HCO_3^-$$

remove

(pull)

(HCl loss from vomiting)

This is because the open system is *pulled* to the right (not pushed). Hence the HCO_3^- generation is nearly unlimited, there being a continuing supply of CO_2 from normal metabolism. In other words, the reaction does not move along the CO_2 dissociation curve but leaves it and moves to another curve. (This concept also finds expression in the Haldane effect. See Fig. 9.) In the present example the reaction moves above the normal curve—a process which can be called ⟨metabolic alkalosis⟩.

Combined Effect of Pco₂ Rise and Fixed Acid Loss on [HCO_3^-]

Suppose a bicarbonate value is high because of *both* a respiratory disturbance such as CO_2 retention and a metabolic disturbance such as vomiting. A "problem" arises. How much of the [HCO_3^-] increase is caused by Pco_2 rise and how much by HCl loss? If a high bicarbonate, e.g. a Δ[HCO_3^-] of, say, +10, is considered apart from the appropriate CO_2 dissociation curve, an exact answer cannot be obtained. To make use of a CO_2 dissociation curve the blood sample in question can be placed in a tonometer,* CO_2 is added until all bases other than HCO_3^- (i.e. Hb^-, $HPO_4^=$ and labelled Buf^-) have been

---

* A tonometer is a vessel "closed" to all materials except gases, i.e. allowing only CO_2 and O_2 to enter and leave the blood sample contained in the vessel. The Astrup micro-tonometer depicted on page 15 is clearly described by Winters *et al.*[23]

returned to their normal concentration (i.e. when pH = 7.4), and a new or corrected $[HCO_3^-]$ and other indices are derived from the CO_2 dissociation curve. This procedure can be actually carried out or, more conveniently, carried out graphically or nomographically (as in Appendix III) and can be used to derive the Base Excess. The procedure would, for example, yield a figure for BE = +8. From this result one could infer that, of the +10 mEq/L of $[HCO_3^-]$ increase, 2 mEq/L of the increase resulted from a P_{CO_2} rise and 8 mEq/L from the vomiting.

Combined Effect of *P*co₂ *Fall and Fixed Acid Accumulation on* [HCO₃⁻]

When fixed acid invades the body (as in diabetic ketosis), hyperventilation often ensues. How much $[HCO_3^-]$ lowering is caused by P_{CO_2} lowering and how much HCO_3^- was used up in buffering the fixed acid? Again, assuming an *in vitro* dissociation curve, this question can be answered by a tonometric method. The meaning of this CO_2 titration method is important to grasp and the method will therefore be restated: lower the P_{CO_2} in the tonometer until $[Buf^-]$ is returned to its state before the acidosis occurred (i.e. until pH is 7.4) and determine (actually or nomographically) a new or corrected bicarbonate (in whole blood).

The principle behind this method is simply that in a closed system (closed except to gases) the *total* base concentration available for buffering fixed acid is *not affected* by P_{CO_2} change. Unfortunately in the blood stream this principle does not apply and hence *in vitro* determinations of "base excess" cannot be fully relied upon for *in vivo* diagnosis. To understand this requires that the chemistry of tonometry be briefly reviewed.

Why "Base Excess" Can Be Misleading

Placing blood in a tonometer closed except to gases endows the blood with a unique capacity to buffer fixed acids, namely, a capacity unchanged by P_{CO_2} changes. This property of blood *in vitro* exists because when CO_2 is added, for example, H^+ is simply transferred between the P_{CO_2}-bicarbonate and Buf systems as follows:

$$
\begin{array}{l}
CO_2 \qquad\qquad\qquad |HCO_3^-| \quad \text{(generated)} \\
\;\downarrow \qquad\qquad\qquad\qquad \uparrow \\
CO_2 + H_2O \rightarrow H^+ + HCO_3^- \\
\qquad\qquad\qquad \downarrow \\
\qquad H\,Buf \leftarrow H^+ + Buf^- \\
\qquad\qquad\qquad\qquad \uparrow \\
\qquad\qquad\qquad |Buf^-| \quad \text{(used up)}
\end{array}
$$

It is clear that, in a system closed except to CO_2, for every mEq of HCO_3^- generated, one mEq of Buf^- is used up in buffering H^+. Thus, in this sytem open only to gases, the total number of anions available for binding H$^+$ ions, i.e. the total number of bases which can act as buffers, is unchanged by the respiratory disturbance of adding CO_2 to the system.

It is therefore possible to change the Pco$_2$ *in vitro* to some other Pco$_2$ without changing the total buffer base, (i.e. the sum [HCO$_3^-$] + [Buf$^-$]). If this is done correctly this sum can be calculated and therefore an estimate of the total amount of fixed acid invasion or loss be derived. Thus can the [HCO$_3^-$] be chemically dissected into a respiratory and metabolic fraction. Thus also can the total buffer base and its deviation from normal (which, by definition, is the "base excess") be determined. The principle underlying this procedure is further explained in Appendix III. It is valid for blood in a tonometer, i.e. blood as a system open in only one way, namely, allowing only the passage of CO_2.

However, the vascular system is open in several ways. It allows the passage of materials other than gases into and out of blood. In particular it is open to the passage of bicarbonate as well as CO_2 because HCO$_3^-$ can migrate in or out of the blood stream through capillary walls. In acute respiratory disturbances, an important example of which is the CO_2 retention of acute respiratory failure, some HCO$_3^-$ ions leave the blood by leaking through the capillary walls to the interstitial fluid. This comes about in the following way. Because of the action of hemoglobin, the blood [HCO$_3^-$] increases *much more rapidly inside than outside the blood stream.* Why is this?

It is essential that the full reason for the blood's ability to generate bicarbonate from CO_2 be explained at this point (even if we are forced to consider one of the interactions between CO_2, O_2 and hemoglobin before coming to the chapter on oxygen). The reason is as follows: the H$^+$ ions in the reaction $CO_2 + H_2O \rightarrow H^+ + HCO_3^-$ are transferred to an especially efficient H$^+$ binder, namely unoxygenated hemoglobin, Hb$^-$, thus pulling the reaction to the right. This efficiency is mainly due to what is called the Christensen-Douglas-Haldane (CDH) effect, or Haldane effect for short. It is illustrated in Figure 9.

As a result of the fact that more HCO$_3^-$ is generated in the blood stream than in the interstitial fluid, in acute CO_2 retention a plasma-to-interstitial fluid [HCO$_3^-$] gradient is established with the result that HCO$_3^-$ ions move into the interstitial space. As can be seen in Figure 10, when Pco$_2$ is acutely raised above 60 mm Hg, the whole body CO_2-dissociation curve is significantly lower than that for isolated whole blood. Such bicarbonate migration would be of no great importance in clinical medicine but for the manner in which it affects certain widely used indices of "metabolic" acid-base disturbance, viz, buffer base and base excess or BE. What happens is that these indices, based as they are on analyses and manipulations of blood *in vitro*, have falsely suggested that a ⟨metabolic⟩ or fixed acid invasion "is intrinsic to acute carbon dioxide accumulation."[42] Discussions of the confusion resulting from the uncritical

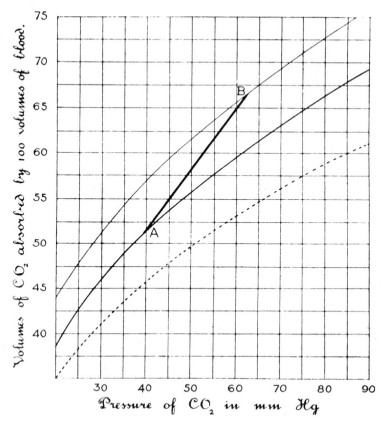

FIG. 9. The Haldane effect. Top curve—venous blood. Middle curve—arterial blood. The steep heavy line represents the physiologic CO_2 dissociation curve of the blood of J. S. Haldane as studied by Christiansen *et al.*[41] Arterial blood (A) avidly absorbs CO_2 because as it loses O_2 it becomes venous blood (at point B) in which hemoglobin is a stronger base (being converted from O_2Hb^- to Hb^-). This stronger base binds protons and thus increases the generation of HCO_3^- (which comprises most of the total CO_2 in blood).

$$CO_2 + H_2O \rightarrow H^+ + HCO_3^-$$
$$\downarrow$$
$$O_2 + HHb \leftarrow H^+ + O_2Hb^-$$

The dotted curve is based on dog and ox blood.

For a recent analysis of the Haldane effect *in vivo* see Michel.[33]

use of such indices have been presented elsewhere.[43,44] The error involved can be stated in one sentence. The largest component of the buffer base is plasma $[HCO_3^-]$; this concentration is lowered when HCO_3^- ions migrate from blood to interstitial fluid; this migration lowers the buffer base in acute CO_2 retention more than expected from nomographic calculations; hence ⟨metabolic

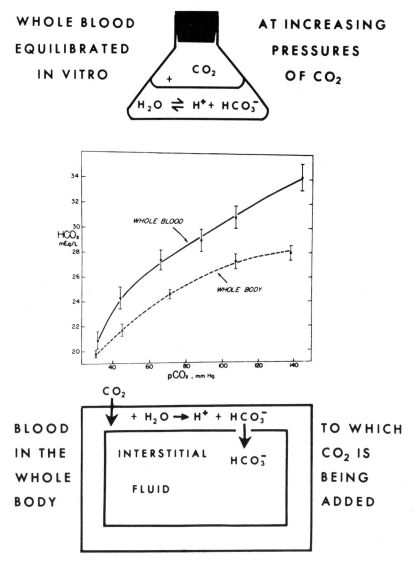

FIG. 10. The CO_2 uptake curves of dog blood in a tonometer and blood circulating in dogs breathing gas mixtures containing 6 to 13% CO_2 for several hours. (Graph reproduced by permission from Cohen et al.[45]) In the P_{CO_2} range of 40 to 60, the *in vivo* slope is about 1:10, half as steep as the curve shown in Figure 7. The slope is 1:7 or 1:8 in man acutely exposed to CO_2.[46]

acidosis⟩ can be wrongly diagnosed from the base excess index and the older "buffer base"[32] from which BE is derived.

The Regulation of the Bicarbonate Base Reserve

Since HCO_3^- is generated from CO_2 and changes with P_{CO_2}, it might be thought that the bicarbonate level in the body is controlled by the lungs. However, it is the fraction of filtered bicarbonate reabsorbed by the kidneys which ultimately determines $[HCO_3^-]$ in plasma. Bicarbonate reabsorption is an almost automatic process in the normal kidney. The total bicarbonate extracellular space (about 15 liters) is effectively filtered in a little over two hours. In this time, nearly the total available HCO_3^- in the body enters the kidney tubules and would be lost were it not reabsorbed. Since less than 3 mEq/L of HCO_3^- ions are lost from the body per day, the kidney reabsorbs virtually 100% of the filtered bicarbonate.

If plasma $[HCO_3^-]$ exceeds the "renal bicarbonate threshold," the tubules simply do not take back the excess over the normal filtrate concentration of 24 to 26 mEq/L.[27] This threshold of 24 to 26 is not fixed but can be varied by the interaction of independent mechanisms.[47,96] What is the HCO_3^- reabsorption mechanism? Since HCO_3^- diffuses across tubular cell walls less easily than CO_2, HCO_3^- reabsorption is generally thought to be brought about by an indirect process—one in which filtered HCO_3^- is first converted to CO_2 which then diffuses from tubular fluid back into renal tubular cells where it is converted again to HCO_3^- and returned to the blood. This process is virtually the same as the one we have been studying all along, at least in the sense that the major factor determining the amount of bicarbonate generation depends on the amount of H^+ transferred between systems. The overall process of HCO_3^- reabsorption can be illustrated by the statement that "the most important feature of the exchange mechanism is the simple quantitative relation that one bicarbonate ion is added to the plasma for each hydrogen ion secreted."[48]

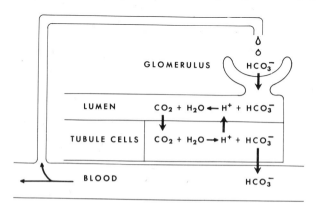

The major difference between renal CO_2 transport and the behavior of the P_{CO_2}-bicarbonate system in the lungs and body tissues is in the postulated H^+ pump.[27] The question whether this pump exists is less important for our purposes than an understanding of the overall process by which plasma $[HCO_3^-]$ is regulated by renal control of the balance between bicarbonate production and excretion, i.e. HCO_3^- balance. This renal control is influenced by five major factors:

1. *The effect of P_{CO_2} changes on renal bicarbonate reabsorption.* When P_{CO_2} is elevated for days, e.g. in severe lung disease or by prolonged inhalation of CO_2 mixtures, the plasma bicarbonate rises.[49] This rise is the result of increased renal tubular reabsorption and is called renal ⟨compensation⟩. It is generally believed, since the work of Brazeau and Gilman,[50] that the stimulus for this increased reabsorption is the high P_{CO_2} of the arterial blood (and not its pH). The ultimate stimulus, of course, may well be the pH change *within* tubular cells since an increase in P_{CO_2} outside these cells drives CO_2 inside them where it can react with H_2O and increase intracellular $[H^+]$.

Much has been written about why, in chronic obstructive pulmonary disease, renal ⟨compensation⟩ is not complete, that is, why the arterial pH is not usually brought to normal. Most recently Robin *et al.*[8] speculate that the arterial pH is not the determining factor but that "the critical compartment where pH is regulated would presumably be renal intracellular water."

2. *The effect of potassium levels on plasma $[HCO_3^-]$.* It is a well-known clinical fact that when plasma $[K^+]$ is low the plasma $[HCO_3^-]$ is high and vice versa. Infusing sodium bicarbonate lowers plasma $[K^+]$. Infusing KCl lowers $[HCO_3^-]$, apparently by two mechanisms: (a) The K^+ seems to enter tissue cells "in exchange" for H^+. This H^+ reacts with bicarbonate and lowers $[HCO_3^-]$, the resulting CO_2 leaving via the lungs.

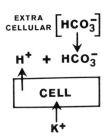

(b) Bicarbonate is lost in the urine (along with K^+) because "as a consequence of reduced intracellular hydrogen ion concentration, fewer hydrogen ions are secreted and less bicarbonate is reabsorbed."[27]

Potassium depletion, an important clinical state, produces the reverse effect and raises $[HCO_3^-]$. When body potassium stores are reduced (as from low K+ intake or by gastrointestinal or renal losses), the plasma bicarbonate is high and the [K+] low—the well-known but poorly understood state of hypokalemic alkalosis. Hydrogen ions apparently enter cells as potassium leaves them and extracellular HCO_3^- is generated from the ever-present CO_2 supply.

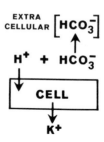

Hydrogen ions also enter tubular fluid in greater than normal quantity producing "paradoxical aciduria," a phenomenon which has received a variety of explanations.[27,51,52]

3. *The reciprocal relationship between plasma* $[HCO_3^-]$ *and* $[Cl^-]$. It has long been known that, when [Cl−] in plasma is low, the plasma $[HCO_3^-]$ tends to be high and *vice versa*. The explanation that the law of electroneutrality requires that this be the case (since a given [Na+] must be "balanced" at least approximately by the sum of $[HCO_3^-]$ and [Cl−]) contributes little to the understanding of acid-base physiology and the significance of anion reabsorption to sodium conservation.

The reabsorbability of Cl− (as contrasted to phosphate and other nonreabsorbable anions) may play a part in HCO_3^- regulation. Thus in *chloride deficiency* there is an increased demand for HCO_3^- to accompany Na+ in the tubular reabsorptive process, a demand which acts as a powerful stimulus to urinary acidification. Bicarbonate "reabsorption" is thus increased and acid is lost in the urine (along with potassium). Repair of the alkalosis associated with hypokalemia does not depend on correcting the K+ deficiency, but can be accomplished simply by replacing chloride by NaCl infusions.[53] Some of the decrease in HCO_3^- reabsorption may result from the effect of volume expansion by such infusions, as discussed below.

4. *The effect of extracellular fluid volume.* It has been known for some time that NaCl infusion can lower HCO_3^- reabsorption.[54,55] That Cl− is not necessary for this effect has recently been shown,[47] since infusion of NaHCO_3 is just as effective as NaCl. It appears that *expansion of the extracellular fluid volume* is an important stimulus to decreased bicarbonate absorption and hence of the level of plasma bicarbonate itself.

5. *HCO_3^- reabsorption and the secretion of adrenal cortical hormone.* Hypersecretion of mineralo- and to a lesser extent glucocorticoids (as in Cushing's

syndrome) and aldosterone tend to elevate [HCO_3^-] levels in the plasma. In Addison's disease the reverse occurs and bicarbonate levels tend to be low. To what extent these effects are secondary to potassium loss and accumulations seen in these disorders is not clear.

MEASUREMENT OF PLASMA [HCO_3^-] AT THE BEDSIDE

Although the traditional Van Slyke vacuum extraction method[56] for [Total CO_2] is still the standard, methods depending on extraction of a quantity of CO_2 (including Natelson's method[57]) are and must always be slow compared to methods depending on intensity factors such as Pco_2 and pH. Furthermore extraction methods are not suitable for bedside use.

The [HCO_3^-] of plasma is rapidly and accurately determined in most modern laboratories by calculation from pH and Pco_2 of blood by the equation

$$[HCO_3^-] = .03 \ Pco_2 \times 10^{pH-6.1}$$

or from a blood gas slide rule[58] or standard nomograms.

THE CLINICAL SIGNIFICANCE OF PLASMA [HCO_3^-]

Why determine [HCO_3^-]? What advantages does it have (1) over the classical CO_2 content and (2) the currently popular "base excess"? Since all three can have the same clinical implication it is important (3) to state this implication clearly.

1. CO₂ Content

The CO_2 content, being normally about twenty nineteenths of [HCO_3^-], is a reasonable approximation of [HCO_3^-]. It has stood the test of time and is deservedly the most widely used index of ⟨metabolic⟩ acid-base disturbances. The CO_2 content of venous blood can be determined in the "central laboratories" of hospitals throughout the world. The following statement is nearly as true today as in 1931:

> "To ascertain whether an outspoken condition of [fixed] alkali excess or deficit exists, it is usually sufficient to determine the bicarbonate *or* CO_2 content of the arterial, or even the venous, blood as drawn."[28] (Italics mine.)

The major difference between the clinical acid-base situation today is that accurate pH meters allow Pco_2 to be measured; hence the frequency of respiratory acid-base disturbances is more generally recognized than in 1931.

The main advantage of measuring and thinking about $[HCO_3^-]$ rather than CO_2 content is not numerical but conceptual. As we have seen, bicarbonate

is a fairly strong base and binds H^+ ions,
forms the major extracellular base reserve,
means something totally different from dissolved or Free CO_2,
means something totally different from CO_2 tension (Pco_2).

Clinically the last phrase is the most important because CO_2 content and CO_2 tension are often confused—to the detriment of patients. This confusion is avoided by remembering that $[HCO_3^-]$ comprises most of CO_2 content. Thus the CO_2 content, the sum $[HCO_3^-] + [Free\ CO_2]$, is numerically as useful as ever, but it is always worthwhile to "think of" CO_2 content as $[HCO_3^-]$—if only because the symbol $[HCO_3^-]$ does not "look like" the expression "CO_2 tension."

2. The Base Excess, (BE)

This is a derived H^+-binding index consisting of $[HCO_3^-]$ "corrected" for Pco_2 change and includes the buffering contributions of hemoglobin and other blood proteins. Its complete derivation is in Appendix III.

Changes in BE as observed clinically are very similar to changes in $[HCO_3^-]$ except when Pco_2 changes are extreme (see Fig. 11). This is due to the fact that most of a BE change is a bicarbonate change and BE is not greatly affected by anything else such as the strength of hemoglobin H^+-binding or ordinary alterations in Pco_2. This has been known for many years. As stated in 1931:

"Changes in plasma bicarbonate content that accompany serious acidosis or alkalosis are gross in comparison with variations due to the degree of oxygenation of the blood or to alterations of CO_2 tension . . ."[28]

More recently it has been written[59] that the $[HCO_3^-]$ is "a measure that gives a *mixture* of two components," that "these can be separated quite easily in the analytical procedure" for determining base excess. The Astrup procedure yields a scale on which to gauge how much of the bicarbonate displacement can be considered to have resulted from the fixed acid change, if the displacement was in fact associated with fixed acid gain or loss.

To determine the base excess requires that the hemoglobin concentration be measured. To determine it accurately, the percentage saturation of hemoglobin with oxygen is also needed along with a graphic calculation which takes account of the effect of Pco_2 change, at least *in vitro*. How important are these refinements in (a) acid-base chemistry, (b) acid-base physiology, and (c) diagnosis and treatment?

(a) "Titratable acid, or base excess, is a valuable parameter for the

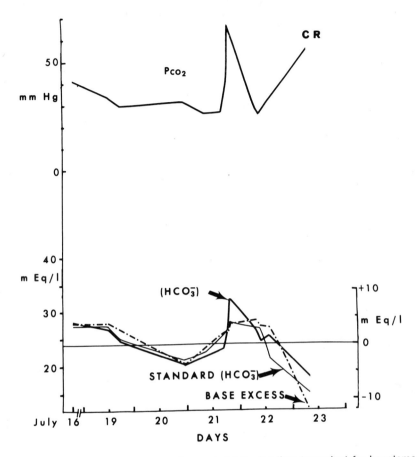

FIG. 11. The patient, in the postoperative period following liver transplant for hepatoma, died on July 23 of pneumonia, atelectasis and gastrointestinal hemorrhage. Acute airway obstruction caused a sudden rise in P_{CO_2}. This is reflected in the rise in $[HCO_3^-]$. Note that neither standard $[HCO_3^-]$ (see Appendix III) nor BE changed very much with this hypercapnia.

description of biological fluids."[60] This is true for the study of *in vitro* systems. BE is also a convenient parameter because "titration with strong acid or base to a pH of 7.40 and a P_{CO_2} of 40 mm Hg at 38°C is difficult as a routine method due to the difficulties in maintaining the P_{CO_2} = 40 during titration."[60] As shown in Appendix III, in a tonometer system closed except to gases, the BE procedure provides an indirect but exact means for determining how *blood* as an isolated system buffers fixed acid and base.

(b) How the *body* buffers fixed acid and base invasion is of greater interest to the physiologist and physician. The open P_{CO_2}-bicarbonate system, because it acts as a vehicle of escape for CO_2, provides the body with an efficient means of disposal of fixed acid; this system is aided by two other bodily mechanisms. As we have mentioned, these are (i) enlistment of the central nervous system which can vary the Pco_2 and (ii) physical transport of H^+ and HCO_3^- between body compartments. The substitution of BE (a derived value) for $[HCO_3^-]$ (a definite chemical compound) would seem to bring no advantage to the physiologist studying these mechanisms.

(c) The BE is used by clinicians both diagnostically and therapeutically. A "minus base excess" is, for example, considered to be evidence of ⟨metabolic acidosis⟩; its magnitude is supposed to indicate how much base (usually sodium bicarbonate) should be injected (usually by intravenous "push"—a term with special significance today) by the formula

$$\text{mEq base needed} = 0.3 \times BE \times \text{body wt in Kg.}$$

This formula is supposed to normalize the acid-base status of the *extracellular space* only.[22]

The diagnostic significance of BE is no different than that of plasma $[HCO_3^-]$ *when other acid-base indices are considered*. As first emphasized by Schwartz and Relman,[61] and agreed to by the proponents of the BE concept, the physician cannot depend on BE alone anymore than $[HCO_3^-]$ alone to assess acid-base disturbances. One must consider the intensity factors Pco_2 and pH as well and especially the clinical circumstances in which the measurements are made. (See Chapters 5 and 6.)

For various reasons the uncritical use of BE in a formula* tends to lead to over-treatment with bicarbonate solutions. Use of the formula encourages "treatment of the BE deviation," rather than treatment of the patient or at least the pH of the patient's blood. As seen in Figure 11, the BE may be higher or lower (more negative) than $([HCO_3^-] - 24)$. This may or may not mean that a greater ⟨metabolic⟩ disturbance exists than is indicated by the change in $[HCO_3^-]$, depending on the distribution of buffers in the various body compartments. A suitable formula, in fact, could be (if one wishes to use a formula),

$$\text{mEq } HCO_3^- \text{ needed} = 0.3 \times ([HCO_3^-] - 24) \times \text{body wt in Kg.}$$

* The above formula was derived from measurements on normal subjects with "diseases unrelated to acid-base metabolism."[64] Most patients with fixed acid loss or gain have disturbances of their electrolyte compartments. The formula is therefore to be considered one of several *very rough* estimates of how much fixed acid needs to be neutralized in patients. The reasons why such estimates must be very rough have been discussed by J. W. Woodbury who summarizes the problem as follows: "On the basis of present knowledge, how much of the excess acid has gone into cells is nearly impossible to assess even if accurate data concerning the time course of the disease are available. Conservative therapy then is to approach but not to try to reach complete compensation."[65]

The practical way to use this or the BE formula is:

> to consider the pH level,
> to calculate how much HCO_3^- to administer,
> to inject half this amount,
> to determine another pH.

These instructions were implied by an important statement in small print in an influential paper:

> "For the *whole body*, the factor 0.7 should be used instead of 0.3 (Palmer and Van Slyke, 1917). In dealing with patients, however, the amount of excess or deficit of base in the whole body does not always seem to be directly proportional to the excess or deficit of base in the blood. When, therefore, patients are to be treated with intravenous infusions of acid or base, it is advisable to estimate the dose necessary to normalize the base content of the extracellular space only, and then follow the effect of the treatment by frequent blood analyses and also by clinical observation, before new infusions are given. Thus overtreatment is avoided."[22]

Other features of the base excess have been widely debated.[44,59-62,72] Unless these are understood by the user, the base excess would seem to have no advantage for clinical purposes over the [HCO₃⁻]. In fact, as discussed in Appendix III, [HCO₃⁻] is probably to be preferred as being a more direct and less confusing measurement.

3. What a Change in [HCO₃⁻] Implies Clinically

It should be remembered, but is often forgotten, that a decrease in plasma [HCO₃⁻] is only *presumptive* evidence that fixed acid has been added to or retained by the body. For example, until the lactate ion has been demonstrated to be in excess, a low bicarbonate level cannot be used to diagnose lactic ⟨acidosis⟩. In fact, until such anions as $SO_4^=$, $HPO_4^=$ and β-hydroxybutyrate anions are actually demonstrated, about all that can be said for sure about a low bicarbonate concentration is that it is low. The same statements hold for "base excess" changes; a "negative base excess" means that the concentration of one of the buffer bases, usually HCO_3^-, is low; fixed acid may or may not have been added to or retained by the body.

Blood, considered *in a tonometer* as a system closed except to the passage of CO_2 gas, can have its bicarbonate lowered in only one way—by the loss of CO_2. While it is true that CO_2 loss often occurs clinically in states of hyperventilation, the consequent lowering of [HCO₃⁻] need not be confusing to a clinician. The amount of [HCO₃⁻] lowering to be "expected" is dealt with in Chapter 5. To trust a reduction in BE as a measure of how much fixed acid has been added to the body is not justified because BE mainly reflects bicarbonate levels and these can change without there being either a ⟨metabolic⟩ or ⟨respiratory⟩ disturbance in the usual sense.

We have discussed how plasma $[HCO_3^-]$ can be lowered by leakage from the blood stream. Here we will give three more illustrations of $[HCO_3^-]$ changes *not* associated with strong acid addition or loss.

a. *Low bicarbonate caused by renal loss of HCO_3^-*.   When carbonic anhydrase inhibitors (e.g. Diamox) interfere with the kidney's daily task of reabsorbing 99% of the filtered bicarbonate, the plasma $[HCO_3^-]$ will become lowered. Several drugs of the sulfanilamide group inhibit carbonic anhydrase and several clinical examples of what is called "metabolic ⟨acidosis⟩" caused by bicarbonate loss in the urine have been documented.[66,67]

b. *Dilution ⟨acidosis⟩*.   The so-called "acidifying effect" of NaCl[68,69] has been a confusing topic for many years. It is now clear that when NaCl solutions are infused intravenously, most of the lowering of pH in the blood (acidemia) is caused by a lowering of $[HCO_3^-]$ in the plasma as a simple result of the *addition of water to extracellular fluid*.[70] The added water lowers the HCO_3^- concentration but does not change the P_{CO_2}; hence the pH falls. (See Fig. 12.)

Such $[HCO_3^-]$ lowering can be labelled a "metabolic acidosis," but both the terms of the label are misleading. It should be called dilution acidemia.

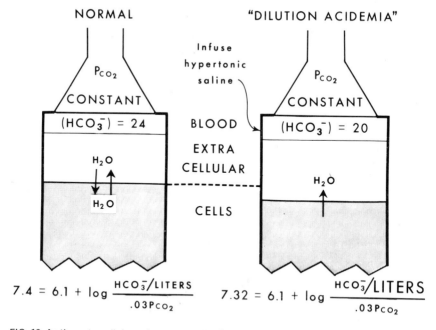

FIG. 12. As the extracellular volume expands, the number of liters increases in the relationship $[HCO_3^-] = HCO_3^-/\text{Liters}$. Such dilution might be thought to *lower* $[H^+]$ as well as $[HCO_3^-]$. This does not occur because, with dilution, carbonic acid is further dissociated,[31] thus producing more H^+.

c. *Contraction alkalemia.* The converse of the foregoing occurs when [HCO$_3^-$] is raised by loss of water, that is, when bicarbonate is concentrated. This can occur after diuresis of edematous patients with agents such as ethacrynic acid. Urinary loss of large volumes of water containing but little bicarbonate causes "the extracellular space to contract around the initial extracellular bicarbonate pool."[71] Hence, the [HCO$_3^-$] rises and the pH must rise as seen in the following equation:

$$pH = pK + \log \frac{HCO_3^-/\text{decreased ECF}}{0.03\ Pco_2}$$

(The numerator of the log ratio increases; hence pH increases.)

These three examples suffice to emphasize the fact that changes in [HCO$_3^-$] do not always indicate that organic or mineral acid or base have invaded the body, and to label all such changes ⟨metabolic⟩ acid-base disturbances simply adds unnecessary confusion. In the words of E. B. Brown,[72] in discussing the lowering of buffer base caused by leakage of HCO$_3^-$ from the blood stream, "What is actually measured then is a bicarbonate deficit; it seems to me that this term has more meaning than the term metabolic acidosis."

In summary, the plasma [HCO$_3^-$] always falls when fixed acid enters the blood, but it cannot always be trusted as an index of fixed acid gain or loss if considered in isolation. Neither can the "base excess." However, *adequately relating* [HCO$_3^-$] *to the* Pco$_2$ *adds significance to both measurements.* Adequately relating these variables includes calculating their ratio which yields the pH. This will be done in the following chapter.

3

....

The Intensity of Acidity, pH

"Many relations implicit in the general equations of acid-base equilibria do not appear vivid and do not find their way into every day practice until they are reargued, reformulated and named."

W. M. CLARK (1928)[31]

....

SINCE blood pH measurement is now available to physicians, and furthermore can be of life-saving importance in therapy, it is more urgent that the mystery surrounding the concept of pH be stripped away than when the above words were written. There seem to be two reasons for this mystery, one superficial and one fundamental. The superficial reason is that pH, although a conveniently sized number, is inconveniently (logarithmically) related to an extremely small magnitude, namely, the concentration of H^+. In most textbooks on acid-base regulation the mathematical connections between pH and $[H^+]$ are displayed in detail to the dismay of virtually all medical students and physicians. These connections deserve to be in a footnote as in Table 3.

The fundamental reason for the mystery is that $[H^+]$ and pH are ways of expressing the *intensity* of acidity and this is a more subtle concept to grasp than a *quantity* of potentially ionizable hydrogen or concentration of acid. Simply stated, the difference is that a small quantity of a strong acid, because it yields more H^+ ions than the same quantity of a weak acid, can produce a more intensely acid solution than a large quantity of a weak acid. Similarly a large quantity of metabolically released H^+, because it is immediately bound by body bases, can be disposed of with little change in body acid intensity.

We will develop means of visualizing $[H^+]$ and pH as indices of acid intensity, describe the process of H^+ transfer, summarize the mechanisms by which potentially ionizable hydrogen, symbolized by H, can be transported and disposed of in large quantities and bring together certain acid-base concepts needed to interpret pH measurements at the bedside.

49

TABLE 3. *"Acidity" has two factors*

	Symbol	Concept	Measure	Unit
Intensity factor	H^+	proton, "hydrogen ion"	pH*	pH unit
Quantity factor	H	potentially ionizable hydrogen	TA	mEq

The expression of "hydrogen ion concentration" in concentration units such as nanomoles per liter is to be avoided because it tends to obscure the distinction between the intensity of acidity, pH, and a quantity of titratable acid, TA. Similar fundamental objections to substituting $[H^+]$ for pH were clearly stated by Clark[31] and have been recently restated.[73-77]

---

* The lower the pH, the higher the acid intensity. A fall in pH of 0.3 is equivalent to a doubling of $[H^+]$. Formally these relations are illustrated by:

$$pH\ 7.4,\ [H^+] = 10^{-7.4} = 10^{0.6} \times 10^{-8} = 4 \times 10^{-8}\ molar$$

$$pH\ 7.1,\ [H^+] = 10^{-7.1} = 10^{0.9} \times 10^{-8} = 8 \times 10^{-8}\ molar$$

Exponential notation is superficially confusing and awkward. The concept of pH as an intensity factor or a chemical potential should not be obscured by the awkwardness of the formalism relating this concept to that of a concentration.

H^+-TRANSFER AND H-TRANSPORT

The Origin and Transient Existence of H^+ Ions in the Body

The waste products of metabolism are mostly acidic in character, meaning that they are acids and therefore contain potentially ionizable hydrogen. The vast bulk of this hydrogen is normally disposed of without difficulty since most metabolism is aerobic and since the main product of aerobic metabolism, by which foodstuffs are completely oxidized, is the volatile "carbonic acid," i.e. CO_2 and water. The overall reaction is:

$$(CH_2O)_6 + 6O_2 \rightarrow 6H_2CO_3 \rightarrow 6CO_2 + 6H_2O.$$

glucose H harmless in water

The proton, H^+, denoted by the acid, H_2CO_3, has but a transient existence. This is because the ionization of an acid (the donation of its protons) involves the transfer of these protons to bases of various strengths. Eventually, as we will see, the original H^+ ions become bound by the strongest base in the body, OH^-, and form H_2O.

Other H^+ ions are not so easily disposed of. A very small fraction (normally 0.2%) of acid production is from incomplete oxidation and from certain protein sources. These acids are called fixed because they are not volatile and (unless oxidizable, like lactic acid, to CO_2 and water) cannot be disposed of as carbonic acid. Examples of these sources of hydrogen ions are:

$$\text{glucose (anaerobically burned)} \rightarrow \text{lactate ions}^- + H^+$$
$$\text{cystine} + O_2 \rightarrow \text{sulfate ions}^- + H^+$$
$$\text{phosphoprotein} + O_2 \rightarrow \text{phosphate ions}^- + H^+$$

In many diseases, either because of anaerobic or otherwise disturbed metabolism or because of failure of renal excretion, these fixed acids, that is, their anions and the H^+ ions they release, accumulate in the body. It is not clear whether it is only the H^+ ions which are harmful in these non-respiratory disturbances, or whether the surplus anions ($SO_4^=$, $HPO_4^=$) are themselves toxic in some way. The overall process whereby H^+ and unwanted ions are disposed of is indicated in Figure 13.

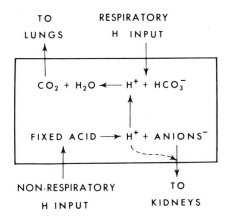

FIG. 13. The major sources and routes of disposal of H^+ ions. The dotted arrow indicates indirect transfer—probably by renal generation of H^+ ions. To dispose of acid means to transfer H^+ to certain bases, thus sequestering potentially ionizable hydrogen (H) into compounds which are either such weak acids as to be harmless to the body (like H_2O), or can be excreted as weak acids from the body (like NH_4^+). The first process involves H^+ transfer, the second H-transport. The distinction was made in 1908 by Henderson.[25] It has recently been elaborated in the chemiosmotic hypothesis of Peter Mitchell[78]: "In the metabolism of living organisms it is possible to distinguish two main types of process. One type of process is described in terms of chemical reactions—the enzyme-catalysed reassortment of atoms and chemical groups joined by primary bonds in the biological polymers and metabolic intermediates. The other type of process is described in terms of transport or osmotic reactions—the permeability-dependent redistribution of inorganic salts and metabolic intermediates across the natural membranes that are responsible for the spatial compartmentation of the chemical reactions. These two types of process have their counterparts in the chemical laboratory or factory and in the thermodynamic formulation of chemical equilibria, where we distinguish the scalar aspect of the chemical processes of primary bond exchange from the vectorial physical transport processes by which the reactants and resultants are taken up to and brought down from the states required at the so-called sites of chemical reaction."

The transient existence of the H^+ ion or proton is clearly described by Edsall and Wyman[6]: "an isolated proton . . . can exist for an appreciable period of time only in a gaseous phase at very low pressures. In any liquid phase, however, a proton reacts, almost as soon as it is liberated, with a neighboring molecule. Thus an acid can act as a proton donor only in the presence of a suitable proton acceptor—that is, another base"—i.e. a base *other* than the (conjugate) base of the acid. This concept of H^+ being transferred from one acid to the base of another acid is insufficiently appreciated as the fundamental reaction of acid-base chemistry, partly because the symbol H^+ is omitted from the usual formulation.

The understanding of this reaction in biologic systems is facilitated by a clear explanation of the meaning of a base since the relative strengths of bases largely determine the intensity of acidity in body fluids.

The fundamental reaction is usually formulated as follows: Let A_1 stand for an acid, and B_1 for its conjugate base. To a solution of A_1 we now add a base, B_2, to which the conjugate acid is A_2. The acid base reaction is then

$$A_1 + B_2 = A_2 + B_1.$$

Even though H^+ exists but transiently and in infinitesimally small concentration, we will write this reaction as

$$A_1 \rightarrow H^+ + B_1$$
$$\downarrow$$
$$A_2 \leftarrow H^+ + B_2,$$

from which it is immediately obvious that H^+ is transferred from the A_1, B_1 system to the A_2, B_2 system. Such transfer occurs because B_2 is a stronger base than B_1.

The Strength of Bases

In these days it is necessary to defend the use of verbal or qualitative definitions, so entrenched is the view that mathematical formulation is all that is needed in science. But exclusive attention to exact numerical relationships can delay progress in understanding; indeed, as Bell has pointed out in considering the progress made in the exact description of equilibria, "the success of these quantitative developments helped to obscure some logical weaknesses in the qualitative definitions."[5] One of the greatest advances in acid-base chemistry in the last 50 years has been a conceptual and qualitative one, namely, the definition of a base as a proton- or H^+-acceptor. Bronsted arrived at this conception from logical considerations which were independent of the numerical relationships between what used to be called acids, bases and salts. He repeatedly used the word "character" in describing the behavior of acid-base variables in 1928:

"Those substances, designated as bases, are, quite generally and independently of the solvent, characterized by their *ability to take up hydrions*. Scheme (1) will consequently express both the acid character of the molecule A and the basic character of the molecule B and can therefore be used as the schematic definition for both."[79]

$$A \rightleftarrows H^+ + B$$
$$Acid \rightleftarrows H^+ + Base \tag{1}$$

In this scheme an acid dissociates into H^+ and a base. The base of a given acid is called the *conjugate* base of that acid. (Thus HCO_3^- is the conjugate base of carbonic acid and $SO_4^=$ is the conjugate base of sulfuric acid.)

Bronsted's definition of a base started a revolution in acid-base thinking which is not yet completely terminated because at least one of its implications is not yet fully appreciated. This is that the strength of a base is as important a concept as the strength of an acid.

Traditional emphasis on acidity and H^+-donation (dissociation) rather than on basicity and H^+ binding or association is partly an historical accident resulting from the way Faraday discovered the process of ionization (electro*lysis*), and Arrhenius formulated his theory of electrolytes. Since Bronsted's day it is logical to attribute as much importance to the conjugate base of an acid as to the acid itself and to focus on the H^+-binding property of a solution as determined by the relative strengths of its bases in developing a quantitative understanding of pH and acidity.

A base is a substance which makes dissociation possible by accepting protons. The most abundant base in the body is water. As such it accepts protons from strong acids. Water is thus more than a solvent. In fact, in Bronsted's (still generally unappreciated) words, "if the solvent completely lacks the character of a base, no 'dissociation' will take place and the acid will dissolve unchanged in the medium . . . the 'dissociation' of an acid, when introduced in a pure solvent, is therefore dependent upon the basic character of the solvent, since 'dissociation' really means nothing else but the transference of the hydrion from the acid to the solvent."[79]

This statement can be illustrated as follows:

$$HCl \rightarrow H^+ + Cl^- \text{ (very weak base)}$$
$$\downarrow$$
$$H_3O^+ \leftarrow H^+ + H_2O \text{ (weak base)}$$

Three common acids stronger than H_3O^+ are HCl, HNO_3 and H_2SO_4. An equivalent statement is that their conjugate bases Cl^-, NO_3^- and $SO_4^=$ are very weak. "In water the 'Big three' are leveled to the strength of H_3O^+."[80] An equivalent statement is that water is a strong enough base to compete successfully for the protons of most strong acids.

However, substances that are stronger bases than water are able to extract protons from H_3O^+:

$$H_3O^+ \rightarrow H^+ + H_2O$$
$$\downarrow$$
$$NH_4^+ \leftarrow H^+ + NH_3 \quad \text{(a base stronger than } H_2O\text{)}.$$

The hydroxyl ion is the strongest base that can exist in aqueous solutions; it is able to extract protons from the ammonium ion:

$$NH_4^+ \rightarrow H^+ + NH_3$$
$$\downarrow$$
$$H_2O \leftarrow H^+ + OH^- \quad \text{(a base stronger than } NH_3\text{)}.$$

It is clear that these reactions are part of a hierarchy. When they are listed in order of increasing base strength (increasing pK value, to be defined), one can visualize protons tending to be extracted successively from proton-donating systems by proton-accepting systems. In Table 4, H^+ tends to move

TABLE 4

Acid			Proton	+		Conjugate Base	pK
Very strong	HCl	$\rightleftharpoons$	H^+	+	Cl^-	Very weak	Negative
Strong	H_3O^+	$\rightleftharpoons$	H^+	+	H_2O	Weak	—
Fairly strong	$HHbO_2$	$\rightleftharpoons$	H^+	+	HbO_2^-	Fairly weak	6.7
Fairly weak	HHb	$\rightleftharpoons$	H^+	+	Hb^-	Fairly strong	7.9
Very weak	NH_4^+	$\rightleftharpoons$	H^+	+	NH_3	Very strong	9.3
Weakest	H_2O	$\rightleftharpoons$	H^+	+	OH^-	Strongest	15.7

Acid-base pairs in order of the strength of their bases. The approximate pK values are from Edsall and Wyman,[6] and Frisell.[82] The first and last pairs listed cannot act as buffer systems; in the range of body pH, Cl^- binds no protons and OH^- binds them so firmly that the resulting acid, H_2O, can only donate a negligible number of protons. pK is the acidic dissociation constant.

The intensity of acidity is directly given by $[H^+]$ which could be called the escaping tendency of protons; it is more convenient[31] to consider the proton-binding tendency of solutions as measured by pH than by $[H^+]$. A modern authoritative statement follows: "In the Bronsted theory, the 'acidity,' an intensive property of a solution, is regarded as the level of proton availability or, in thermodynamic terms, the 'proton activity.' Underlying this definition is the idea that the solute-solvent system consists of several acid-base conjugate pairs. The strengths of the acidic species differ, as indicated by differences in their acidic dissociation constants; in other words, the several bases differ in the tenacity with which they can hold protons."[83]

from the top to the bottom of the list. This process of proton transfer always involves water as a proton carrier in mammalian acid-base reactions.*

The strength of a base, A^-, that is, its ability to extract protons from the acids of other systems, is directly proportional to the pK of the conjugate acid-base pair (HA, A^-). In qualitative terms this means that the stronger a base and the higher its concentration in relation to its conjugate acid, the more protons it can extract from other systems (including water).

It is more important to consider the properties of an acid-base system (e.g. a solution) than those of a single component. The acid intensity of a solution is inversely related to the strength and concentration of its bases, i.e. to the *proton-binding tendency* of the solution as measured by its pH. This is expressed in the Henderson-Hasselbalch equation which we are about to derive; a verbal statement of this equation is:

Proton-binding tendency of a solution	is proportional to	Strengths of its bases plus Ratios of the concentrations of its bases to those of their conjugate acids

A quantitative description of a system of acid-base pairs requires that it be at equilibrium. Usually only one Henderson-Hasselbalch equation is used with the implication that only one acid-base pair is involved. It is important to derive equations describing two buffer pairs at equilibrium,

$$HA_1 \rightleftarrows H^+ + A_1^- \quad \text{(weak base)}$$
$$\updownarrow$$
$$HA_2 \rightleftarrows H^+ + A_2^- \quad \text{(stronger base)}$$

where the vertical arrows indicate that H^+ is being shared, or transferred, between pairs at equal rates, and the horizontal arrows indicate that the rate of proton-binding by the bases equals the rate of dissociation of their respective conjugate acids.

Consider the rate, v_a, of the dissociation of HA_1. Clearly this is proportional to the concentration of HA_1:

$$v_a = k_a[HA_1].$$

* The proton transfers in the reaction $HCl + NH_4OH \rightarrow NH_4Cl + HOH$ can be written in the notation devised[81] for depicting electron carrier mechanisms:

The velocity of the reverse reaction whereby HA_1 is regenerated by A_1^- binding H^+ depends on the concentrations of both A_1^- and H^+. Why?

Suppose, at one instant, an equal number of H^+ and A_1^- ions. Then imagine that the number of H^+ was tripled. The number of collisions with A_1^- would be tripled. Now suppose that the original tripling was not of H^+ but of A_1^- ions. Again, for the same statistical reason, the number of collisions would be tripled. But if the tripling of both H^+ and A_1^- took place simultaneously, the number of collisions would be nine times the original. It is the *product* of the concentrations which determines the number of collisions. Hence the velocity of proton-binding, v_b, will be

$$v_b = k_b[H^+][A_1^-].$$

At equilibrium $v_a = v_b$ and therefore

$$\frac{[HA_1]}{[H^+][A_1^-]} = \frac{k_b}{k_a} = \frac{1}{K_1}$$

where K_1 is the usual acid dissociation constant (often written K_a' to indicate that these relations are only approximate).

Because the normal value of $[H^+]$ is of the order of $10^{-7.4}$ moles per liter of extracellular body fluid, it is convenient to rewrite the last equation as

$$\frac{1}{[H^+]} = \frac{1}{K_1} \times \frac{[A_1^-]}{[HA_1]}.$$

Entirely similar considerations of the behavior of the stronger base A_2^- leads to the relation

$$\frac{1}{[H^+]} = \frac{1}{K_2} \times \frac{[A_2^-]}{[HA_2]}.$$

By practical definition, $pH = \log(1/[H^+])$ and $pK = \log(1/K)$; since only one $[H^+]$ can exist in a solution at equilibrium, the last two relations can be equated and expressed logarithmically:

$$pH = pK_1 + \log\frac{[A_1^-]}{[HA_1]} = pK_2 + \log\frac{[A_2^-]}{[HA_2]}.$$

The equation simply states that H^+ has been bound (mopped up) by the bases A_1^- and A_2^- and that at equilibrium bound H is shared by (distributed between) the acids HA_1 and HA_2.

Chemical Buffering in a Closed System

We will now show quantitatively how bases of two strengths, labelled A_1^- and A_2^-, can share in mopping up added H^+ ions and thus absorb large quantities of acid with little change in acid intensity.

The quantitative description of a biologic buffer system at equilibrium must include three facts:

free H+ ions   many are bound   more are bound
are scarce   by weak bases   by strong bases

$$pH = pK_1 + \log \frac{[A_1^-]}{[HA_1]} = pK_2 + \log \frac{[A_2^-]}{[HA_2]}$$

When this equilibrium is disturbed, say by adding H+ ions, the change in acid intensity (pH) depends on how the added hydrogen is distributed between the two buffer systems. This in turn depends largely on whether the total system is closed or open. We will first consider the system closed as in Figure 14.

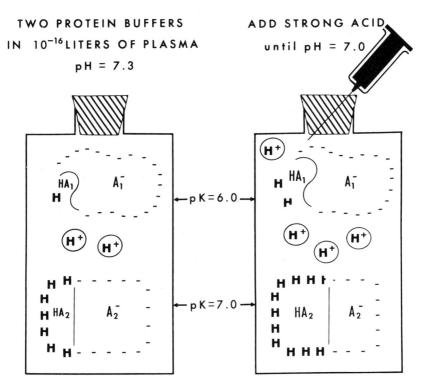

FIG. 14. Buffering in a closed system illustrated by two proteins at the same concentration, i.e. [Protein$_1$] = [HA$_1$] + [A$_1^-$] = [Protein$_2$] = [HA$_2$] + [A$_2^-$]. This concentration is misleadingly low in the drawing, there being actually 20,000 protein molecules per H+.

The total number of H+-binding sites is the same for each protein, i.e. 21. Because their pK's are different, the proteins have different proportions of their binding sites occupied by hydrogen at a given pH.

The higher the pK, the more sites are occupied at a given pH; the lower the pH, the more sites are occupied at a given pK in this example.

It may be seen that most of the added H^+ ions become bound by the stronger base A_2^-, many are bound by A_1^- and very few H^+ ions remain in solution. The intensity of acidity doubles (because $[H^+]$ doubles) *but this is not a measure of the quantity of acid added.*

The quantity of acid added is measured by the increase in the sum $[HA_1]$ plus $[HA_2]$, i.e. the titratable acidity. In a closed system this sum is virtually equal to the decrease in the sum: $[A_1^-]$ plus $[A_2^-]$—in other words, the decrease in the total buffer base concentration. The pH of a body fluid gives almost no indication of its titratable acidity because the two are not linearly related. The exact connection between them is usually shown in titration curves. For our purposes, however, the visual picture in Figure 14 and two simple Henderson-Hasselbalch equations are more important since they display the difference between acid quantity and acid intensity in a closed system reaction.

Before studying the equations to follow, it may be helpful to *count* the H's and binding sites (minus signs) in Figure 14, for they represent the quantities in the equations. Also it may be helpful to recall that $\log 20 = 1.3$, $\log 2 = 0.3$, $\log 10 = 1$, and $\log 1 = 0$.

Before adding acid to the system in Figure 14, the equilibrium is

$$7.3 = 6.0 + \log \frac{20}{1} = 7.0 + \log \frac{14}{7}.$$

from which $[A_1^-]:[HA_1] = 20:1$ and $[A_2^-]:[HA_2] = 2:1$ as shown in the figure. After adding acid, the equilibrium is

$$7.0 = 6.0 + \log \frac{19}{1.9} = 7.0 + \log \frac{10.5}{10.5}.$$

Comparing these two sets of equations it can be seen that lowering pH 0.3 units halved the ratio of $[A^-]$ to $[HA]$ for both base-acid pairs. However, the stronger base, A_2^-, bound more H^+ ions than A_1^- as shown by the fact that its concentration dropped from 14 to 10.5. (The A_2^-, HA_2 pair started with more H's as seen in the figure, and acquired more than the other pair when acid was added.) This is an example of the fact that H^+ binding is especially effective if a base (here A_2^-) has a pK near the pH of the solution. Equivalent for equivalent, the base with the pK closer to the pH is thus the better buffer—in a closed system.

Chemical Buffering in an Open System

It is often said that since the overall pK of the P_{CO_2}-bicarbonate system in plasma is 6.1, it is not a particularly effective buffer pair in bloods with pH's in the neighborhood of 7.4. This is only true in a closed system. We will now show by calculation what was shown qualitatively in the previous chapter (page

28), namely, that the maintenance of a nearly constant Pco_2 in one buffer pair and the process of H^+-transfer between this and other buffer pairs in extra- and intracellular fluids "surpass the efficiency of any possible closed aqueous solutions of like concentration in preserving hydrogen ion concentration."[25]

Because of its historical importance, we will use a calculation made by Henderson[25] for 0.1 M bicarbonate and 0.1 M phosphate, calling the latter Buf.* The modern form of the statement that at equilibrium H^+ ions are "shared" between the two systems is that both systems must be at the same pH.

$$pH = pK_{HCO_3^-} + \log \frac{[HCO_3^-]}{0.03\ Pco_2} = pK_{Buf^-} + \log \frac{[Buf^-]}{[H\ Buf]}$$

Before adding acid

$$7.30 = 6.42 + \log \frac{0.088}{0.012} = 6.60 + \log \frac{0.083}{0.017}$$

After adding acid

$$7.00 = 6.42 + \log \frac{0.088 - x}{0.012} = 6.60 + \log \frac{0.083 - y}{0.017 + y}$$

At the new equilibrium, HCO_3^- must have bound x moles of H^+ while Buf-bound only y moles. Calculation shows that x = 0.042 and y = 0.011, from which we conclude that the Pco_2-bicarbonate system binds nearly four times more H^+ than the Buf system. The arithmetic shows that this is "because Pco_2 is constant" (see denominators of the third equation). In other words, the superior buffering capacity of HCO_3^- (despite its lower pK) is a simple consequence of the fact that when two ratios, the first with a constant denominator, the second with a constant sum of numerator plus denominator, change proportionally (as they must with any pH change), the numerator of the first ratio must change more than the numerator of the second. In other words, x must be larger than y. (See Fig. 15.)

The arithmetic merely quantitates what was illustrated in the chemical formulation of page 28 of Chapter 2, namely, that the constancy or near constancy of Pco_2 allows the transfer of a large quantity of H^+ ions, i.e. $0.042 + 0.011 = 0.053$ moles. This quantity is, of course, almost exactly

---

* In the 60 years since this calculation was made, the pH convention and accurate values for the pK of CO_2 and phosphate systems have been established. Such refinements, as L.J.H. recognized,[3] are "not very important in a discussion of this question, for the general form of the functions involved in the equilibrium is more to the point than the exact value of certain variables." This lack of concern for exactness is perhaps to be found in the classic 1908 paper in which, on page 432, an error appears: 0.082 should read 0.088.

ENHANCED H⁺-BINDING BY AN OPEN SYSTEM

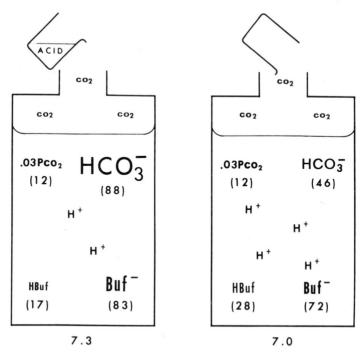

FIG. 15. Buffering in an open system such that the Pco_2 in the gas phase (and hence in the liquid phase at equilibrium) is constant. The numbers in parentheses are from the calculations of Henderson,[25] converted from moles to millimoles.

equal to the amount of H^+ added because the concentration of unbound protons both before and after the addition of acid is of negligible magnitude. (The sum, 0.053 moles, is therefore the *total change in "buffer base,"* that is, the "base excess"; see Appendix III.)

Thus even without the hyperventilation which occurs clinically with accumulation of fixed acid in the body and which will be discussed in a later chapter, the fact that CO_2 can escape from an open system "by a process which under ordinary circumstances has all the appearances of a simple physical phenomenon"[2] tends to minimize changes in the acid intensity (pH) of the extracellular fluid despite the addition of a considerable quantity of acid.

The Immediate and Eventual Disposal of H+

The immediate disposal of H^+ is by *sequestration*, as H in weak acids by the process of buffering which we have just reviewed.

The eventual disposal of most H^+ produced by metabolism is by *incorpora-*

The Intensity of Acidity, pH   61

tion, as potentially ionizable H in water. The disposal of H^+ formed by fixed acid production is by renal *excretion*.

It is often stated that the eventual elimination of ⟨metabolic⟩ fixed acid must be by the kidneys, but what is meant by this statement is not clear. A typical fixed acid is lactic acid, formed from anaerobic metabolism. It is true that lactate, $SO_4^=$, and similar anions are excreted by the kidney but "the excretion of the anions usually presents little difficulty (unless the rate of glomerular filtration is grossly diminished) for they exist free in the plasma and enter the glomerular filtrate; all that is necessary is for the tubules to refrain from reabsorbing them."[48]

What happens to the H^+ ions associated with fixed acids? Is there something different about these H^+ ions that they should be called "metabolic"? No, because as soon as they are formed they, like any other H^+ ions, are immediately bound by HCO_3^- or protein buffers and become either tied up in H_2O via the P_{CO_2}-bicarbonate system or temporarily stored as H Buf. How then can the kidneys excrete the H^+ of fixed acid when this H^+ has been either trapped in body water or sequestered in the body proteins? By generating new hydrogen ions and secreting them into the tubular lumen by the process explained in the previous chapter and illustrated in Figure 10 but with an important difference. Instead of virtually all H^+ reacting with HCO_3^-, some of it is bound by $HPO_4^=$ or NH_3; in other words, some H^+ is trapped in the urine and leaves the body as titratable acid ($H_2PO_4^-$) and ammonium ion (NH_4^+).

This process is depicted in Figure 16. As can be seen, the quantity of non-respiratory H^+ or acid normally excreted is only about 1% (50/5000) of the kidney's total contribution to acidity regulation; 99% of this contribution is in maintaining the bicarbonate reserve by restoring to the plasma the HCO_3^- which would be otherwise lost in the urine. Indeed the end result of even the 1% is simply the reabsorption of additional bicarbonate. "For every hydrogen ion excreted as titratable acid in the urine, an additional bicarbonate ion is added to plasma over and above those bicarbonate ions which merely replace those lost in glomerular filtration."[48] This 50 mEq of HCO_3^- reabsorbed daily is all that is needed to regenerate the non-bicarbonate buffers which have sequestered the 50 mEq of H produced by anaerobic or sulfur metabolism. "Thus the kidneys finally regenerate the body's buffers without ever excreting the actual H^+ which were responsible for their depletion. The carbon dioxide derived from carbonic acid produced by the reactions of bicarbonate with various buffer acids is removed by the lungs; the former H^+ remain harmlessly behind as hydrogen atoms in molecules of the body water."[85]

"Acid-Base Balance" versus pH Regulation

The term ⟨acid-base balance⟩ has almost as many meanings as ⟨acidosis⟩. (See Appendix I.) It should never be used to indicate the base:acid ratio of

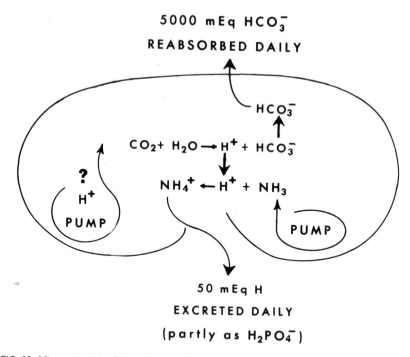

5000 mEq HCO₃⁻

REABSORBED DAILY

50 mEq H

EXCRETED DAILY

(partly as H₂PO₄⁻)

FIG. 16. After secreted H^+ has destroyed filtered HCO_3^- and CO_2 has entered tubular cells, new HCO_3^- is generated in these cells (for reabsorption) and more H^+ is secreted into the tubular lumen to destroy more HCO_3^- as in a previous illustration (p. 38). About 1/100 of the secreted H^+ normally leaves the body as titratable acid or as ammonia.

In non-respiratory acidosis, ammonia assumes great importance. "Ammonia is the major buffer of the urine. Its excretion in acid urine in the cationic form accomplishes the elimination of two-thirds or more of the protons generated within the body in the oxidation of the phosphorus and sulfur of dietary lipids and proteins. If acid production increases in consequence of incomplete oxidations, yielding ketone bodies, the excretion of ammonia increases nearly in proportion to the acid load."[84]

a body fluid because of the fundamental difference between a balance and a ratio. The base:acid ratio of a body fluid is given by its pH, which is the *ratio of two quantities* and therefore an intensity factor.[80] Acid-base balance should denote *the difference between two quantities,* or, more explicitly, the difference between the rate of production or absorption of total acid or base and the rate of its disposal by neutralization or elimination.

Unfortunately there is disagreement on how total or net acid balance should be defined, let alone measured.[48,60,86–91] Siggaard-Andersen wrote that "the concept 'total acid' of a complex buffer solution has no meaning" and that "exact methods for determining the hydrogen ion *balance* of the organism under all conditions are not available."[60] Robinson, a particularly clear writer, wishes "to abandon familiar usage and to speak of 'regulation of the reaction of body fluids' rather than of 'anion-cation' or 'acid-base' balance."[48] Camein *et al.* argue that the term "acid-base balance" be replaced by two other terms: "acidity regulation" and "electrolyte balance."[91]

A generally acceptable concept of acid-base balance is depicted in Figure 17. The contrast between this concept and that of pH has practical as well as fundamental significance. Acid-base balance cannot be easily measured; the pH can be. In fact, short of carrying out actual balance studies[87-91] the H input is largely an unknown quantity.*

For the understanding and evaluation of acid-base physiology in health and disease, an appreciation of buffering by the Pco_2-bicarbonate system and of the automatic processes by which the lungs and kidneys participate in the defence of body pH is more generally important than either balance studies or hints as to how the renal H^+ pump operates.[92] It does not seem to be adequately appreciated how effectively body pH is controlled by simple physicochemical mechanisms operating through open systems. The lung is not the only such open system—the kidney is open to even more materials. Hills and Reid[93] are among the few renal physiologists who recognize that the kidney regulates acid and alkali excretion largely by a process which "exploits the physical and chemical properties of the gases CO_2 and NH_3." The unique fitness of these gases for this process has also been clearly shown.[94] Robinson[48] is among the few who have emphasized the automaticity of pulmonary and renal mechanisms for pH control. In these days when active transport by cellular pumps are almost casually postulated to explain the movement of material through the body, it is of practical importance to consider how much H^+ transport is accomplished by virtue of the structure of the lungs and kidneys and the physical chemistry of the body fluids amenable to direct analysis, since the function of these organs and the composition of these fluids can be changed therapeutically.

THE AUTOMATIC CONTROL OF BLOOD pH BY LUNGS AND KIDNEYS

The body's defenses against blood pH changes operate at different time rates. Chemical buffering is almost instantaneous; pulmonary responses occur

---

* A quantity factor which can be measured is the "net metabolic hydrogen excretion," namely, the sum of urinary titratable acid plus ammonia minus urinary bicarbonate. This measurement is useful in detecting renal tubular acidosis. "The crucial diagnostic test is the response of the kidney to an exogenous load of acid in the form of ammonium chloride under standardized conditions."[95] The importance of the type of exogenous anion load is still in doubt.[96]

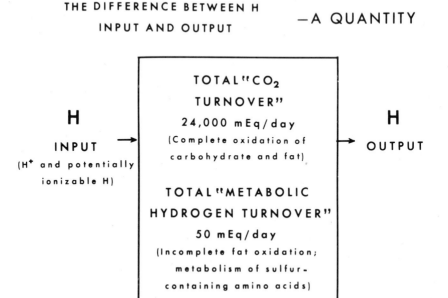

ACID–BASE BALANCE
(applies to whole body)

THE DIFFERENCE BETWEEN H INPUT AND OUTPUT   **—A QUANTITY**

H INPUT
(H$^+$ and potentially ionizable H)

TOTAL "CO$_2$ TURNOVER"
24,000 mEq/day
(Complete oxidation of carbohydrate and fat)

TOTAL "METABOLIC HYDROGEN TURNOVER"
50 mEq/day
(Incomplete fat oxidation; metabolism of sulfur-containing amino acids)

H OUTPUT

BASE:ACID RATIO

BASE CONCENTRATION DIVIDED BY ACID CONC.   **—AN INTENSITY**

$$pH = pK + \log \frac{(BASE)}{(ACID)}$$

(applies to a homogeneous fluid; determines pH for a given pK)

FIG. 17. The contrast between the concept of a balance and a ratio.

in minutes; renal responses in hours to days. Control of pH—especially in disease—cannot be understood without considering the temporal aspect of these various defense mechanisms. No single measurement or set of measurements made at a single point in time—regardless of how elaborately they are correlated or plotted on an acid-base nomogram—can properly depict the acid-base state even of blood. This was recognized by Siggaard-Andersen[62] who in discussing the many possible graphic systems for illustrating acid-base status wrote: "The disadvantage of all these systems is that the time is not illustrated."

To take account of this time factor we will illustrate the *variations* in arterial Pco_2 and in $[HCO_3^-]$ as they affect pH by plotting all three on the classical Pco_2-bicarbonate diagram.

To relate these changes to what we have discussed in the previous chapters, we re-emphasize that pH is proportional to the ratio $[HCO_3^-]/Pco_2$. When this ratio is constant, the pH is constant; when the ratio is high the pH is high. These important concepts are easily visualized in the following diagram:

$$pH \propto \frac{[HCO_3^-]}{Pco_2}$$

$$7.4 \propto log\frac{20}{1} = log\frac{24}{.03 \times 40}$$

The steeper the slope of a line expressing a constant $[HCO_3^-]/Pco_2$ ratio, the higher is the pH which this line represents. The normal CO_2 dissociation curve is also shown passing, as it normally does, through the arterial point defining the normal acid-base state: pH = 7.40, pCO_2 = 40, $[HCO_3^-]$ = 24.

1. Effects of Arterial Pco₂ Changes on pH Control Mechanisms

a. *Pulmonary responses to Pco₂ change.* The rate of CO_2 production is a most important determinant of the rate of CO_2 excretion by the lungs. When CO_2 production increases, CO_2 excretion almost immediately increases also. This is because of the rapidly responding feedback chemoreceptor system discussed on page 8. Augmented ventilation (which normally follows even a few mm Hg rise in arterial Pco_2) immediately increases CO_2 excretion until Pco_2 returns to normal— if the lungs are normal.

A decreased P_{CO_2} (as after prolonged hyperventilation, voluntary or psychogenic) lowers the chemoreceptor CO_2 stimulus and a reduction in ventilation allows dissolved or free CO_2 to build up, restoring P_{CO_2} to normal.

The most serious respiratory disturbance of the acid-base state is CO_2 retention as a result of respiratory failure. This is called respiratory acidosis because even slight increases in P_{CO_2} lower the pH as reflected by the position of the CO_2 dissociation curve. Even if the disturbance "follows the curve" (as shown on page 32 for blood in a tonometer), the pH falls; if the P_{CO_2} rise is very acute, the state leaves the curve (dotted line) and the pH falls even more, partly because $[HCO_3^-]$ leaks out of the plasma as we have shown in Figure 10.

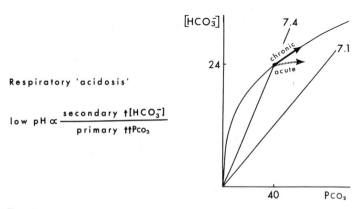

b. *Renal responses to P_{CO_2} change.* When pulmonary disease, then, interferes with CO_2 removal, the P_{CO_2} increases; a small respiratory rise in $[HCO_3^-]$ occurs but not in proportion to the P_{CO_2} rise. Hence pH falls temporarily. In hours to days, however, $[HCO_3^-]$ (and hence pH) begins to rise because the kidneys begin to respond by increased HCO_3^- reabsorption—a so-called compensatory response. This renal response occurs in the following way.

High P_{CO_2} in arterial blood produces an increase in H^+ secretion by the tubule cells.[27] Were it not for this increased secretion, the extra bicarbonate being filtered (because of the respiratory $[HCO_3^-]$ rise in the plasma) would pass down the tubular lumen, resulting in bicarbonate loss and an alkaline urine. However, H^+ secretion exceeds HCO_3^- filtration and for every H^+ which does not react with HCO_3^- and reaches the collecting ducts in TA or NH_4^+, one HCO_3^- is generated in the tubule cell and subsequently enters the plasma. Most of this secondary renal reabsorption of bicarbonate resulting from a primary rise in P_{CO_2} occurs in 48 hours. The direction of this response is shown in the right half of Figure 18; its effect is obviously to restore the blood pH toward normal. In chronic pulmonary disease this "renal compensation" may be complete, i.e. arterial pH may be 7.4.

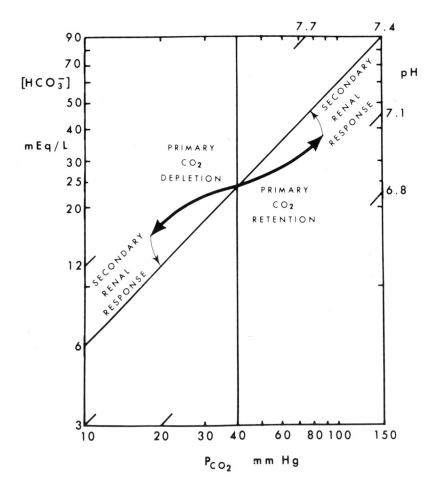

FIG. 18. The two classical "primary respiratory disturbances" causing high or low Pco₂. The secondary (often called compensatory) responses are shown by the light arrows which are drawn as if "complete compensation" (i.e. normal blood pH) were always achieved.

The logarithmic diagram was favored by Peters and Van Slyke[28] for clinical use and more recently by Elkinton.[97] It is similar to the classical linear coordinate system for displaying the CO_2 dissociation curve, but the logarithmic axes expand the lower [HCO_3^-] and Pco₂ scales and therefore are more convenient for plotting clinical data. It is also helpful in research directed at clarifying differences between *in vivo* and *in vitro* curves.[98]

Low Pco₂, such as occurs in normal subjects on arrival at moderately high altitude (5000 to 10,000 feet), produces reverse effects. The secretion of H⁺ by the tubules falls below that of HCO_3^- filtration; bicarbonate automatically passes through the tubules and enters the urine; in several days of acclimatization a definite decrease in plasma [HCO_3^-] will have occurred and the pH, initially high, will eventually be brought to normal. This secondary renal response is depicted on the left side of Figure 18.

2. Effects of Plasma [HCO₃⁻] Changes

a. *Renal responses to bicarbonate change.* The rate of bicarbonate filtration is the most important determinant of the amount of acid excreted in the urine. If HCO_3^- filtration lags behind H^+ secretion, more H^+ ions are free to react with $HPO_4^=$ and NH_3 and to be excreted in the urine. It may well be that the 500 mEq or more[27] of TA and NH_4^+ which daily appear in the urine in such states as diabetic ketosis result from this simple relationship. On the other hand, an increased plasma [HCO₃⁻] automatically increases bicarbonate filtration so that fewer H^+ ions reach the urine. The automatic character of this process is well described by Hills and Reid:[93] "At a relatively fixed rate of H^+ secretion into the nephrons, a small rise of plasma [HCO₃⁻] will soon cause the rate of HCO_3^- filtration to exceed the rate of H^+ secretion; and thereafter further increments of filtered HCO_3^- will not be converted to CO_2 and be reabsorbed, but instead will simply be excreted . . ."

The most serious non-respiratory acid-base disturbance is the destruction of part of the bicarbonate base reserve. Usually the primary *event* is the invasion of fixed acid, but the primary *change* to focus on in non-respiratory disturbances is a low [HCO₃⁻] regardless of how it is brought about:

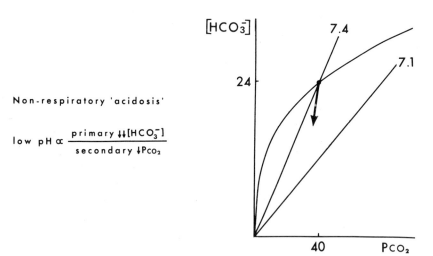

Non-respiratory 'acidosis'

$$\text{low pH} \propto \frac{\text{primary} \downdownarrows [\text{HCO}_3^-]}{\text{secondary} \downarrow \text{Pco}_2}$$

The pH must fall with lowered [HCO₃⁻] unless the Pco₂ falls in exact proportion. Generally it does not do so and the acid-base state "leaves the normal CO_2 dissociation" curve and moves across pH lines toward lower and lower pH values.

b. *Pulmonary responses to bicarbonate change.* The low plasma bicarbonate resulting from HCO_3^- destruction by fixed acid is always associated with hyperventilation, e.g. Kussmaul's air-hunger in diabetic coma. Such hyperventilation lowers P_{CO_2} and thus prevents severe decrease in pH (see lower part of Fig. 19). Gilman has pointed out the interesting paradox that renal control of pH might be interfered with by respiratory lowering of P_{CO_2} "were this to impair the acidification of the urine."[99] It is probably true that hyperventilation delays the eventual restoration of body buffers because low P_{CO_2} inhibits renal H^+ secretion and HCO_3^- resorption. But "the longer time required to excrete all the acid may be a small price to pay"[48] so long as plasma pH is kept under control until the acid invasion is ameliorated.

High plasma $[HCO_3^-]$ may or may not produce hypoventilation, depending on whether or not "as a consequence of intracellular potassium loss, hydrogen ions shift from extracellular fluid into cells."[100] Hypoventilation and P_{CO_2} rise apparently do not occur when intracellular pH falls as a result of K^+ loss (in thiazide diuresis); respiratory adjustment and CO_2 retention do occur in response to the high $[HCO_3^-]$ and high pH induced by bicarbonate infusion or ethacrynic acid as in the upper part of Figure 19.

THE MEASUREMENT OF pH AT THE BEDSIDE

Modern glass pH electrodes if properly cared for and calibrated will yield pH readings in blood accurate to 0.01 pH unit in routine clinical use. A determination takes less than a minute. A part-time technician can be taught to keep the pH meter in working order.

THE CLINICAL SIGNIFICANCE OF BLOOD pH

The following will be but an introduction to the clinical use of pH measurements in these days when (at last) accurate pH meters are now available to and used by physicians caring for patients with such diverse conditions as myocardial infarction, surgical shock and overwhelming bacterial infections—none of which traditionally has been considered an acid-base disturbance. As in the case of P_{CO_2}, arterial is preferable to any venous sample, especially in critically ill patients.

The arterial blood pH, although it is a more important indicator of the *severity* and *acuteness* of an acid-base disturbance than either P_{CO_2} or plasma $[HCO_3^-]$, is less specific than either of these other two measurements. Because

$$\text{pH is proportional to the ratio } \frac{[HCO_3^-]}{P_{CO_2}},$$

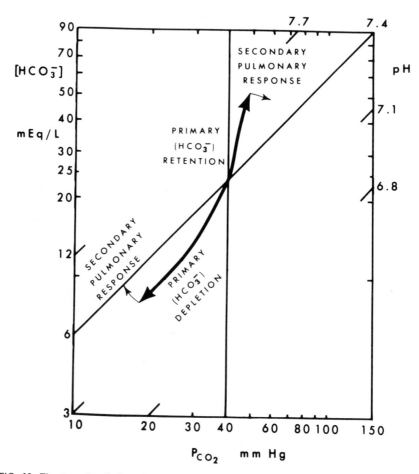

FIG. 19. The two classical "primary non-respiratory disturbances" causing low or high $[HCO_3^-]$. One secondary response is drawn as if complete compensation were achieved.

low pH can result from the numerator $[HCO_3^-]$ being low or the denominator high (or both); high pH can result from a high numerator or a low denominator (or both); normal pH can result from both numerator and denominator being proportionally high or low.

An isolated pH measurement outside "normal" limits (7.36 to 7.44) is thus a sign of significant acid-base disturbance, but in itself gives no indication of the amounts of acid or base retention or depletion. (Again we see that pH is a measure of the intensity of acidity, but not a measure of a quantity of acid.)

When considered in relation to changes in $[HCO_3^-]$ and P_{CO_2}, the pH acquires diagnostic and some* prognostic significance by indicating answers to the following questions:

> How severe is the change in $[HCO_3^-]$ or Pco_2 or both?
> How long have these changes been present?
> Is the $[HCO_3^-]$ or the Pco_2 change the primary change?

1. *Severity of Disturbances*

There is little evidence that either $[HCO_3^-]$ or Pco_2 changes produce toxic effects *in themselves*. It is only as their ratio changes that severe interference with bodily function (especially enzyme function) occurs. For example, a bicarbonate as low as 5 mEq/L (often seen in diabetic ketosis) is not toxic in itself; since the Pco_2 is usually extremely low as well, the pH drop is minimized; patients with such fixed acid invasions often recover completely. In fact, the prognosis in diabetic coma is primarily dependent on the cardiovascular system and the level of consciousness (the latter being a function of brain pH and not of $[HCO_3^-]$ or Pco_2 alone).

Similarly a Pco_2 of 60 mm Hg, so long as pH is not below about 7.20, is not in itself a cause for alarm. (The resultant low alveolar *oxygen* pressure and consequent low arterial oxygen content are probably more significant in respiratory failure than any of the usual acid-base indices.)

2. *Duration of Disturbance*

How long a patient has been ill can often be determined better from clinical observation than from laboratory measurements. However, large deviations of pH from normal suggest an acute process. This is because many secondary adjustments (carried out by the lung and kidney for the most part) take time to be effective in returning the pH toward normal. The paths taken by these "compensatory" adjustments vary in different diseases. Attempts to define these with words and to invent labels for even the more straight-forward of these adjustments has led to much of the confusion of this field.

If a pH is not far from normal and $[HCO_3^-]$ or Pco_2 deviations are considerable, the disturbance is more likely to be chronic than acute.

3. *Is the $[HCO_3^-]$ or the Pco_2 Change Primary?*

Since $[HCO_3^-]$ is affected by fixed acid gain or loss (e.g. in nephritis, and in vomiting respectively), a laboratory report that $[HCO_3^-]$ is, say, 15 mEq/L in a patient with renal disease need not include note of a pH deter-

---

* A pH of 7.0 resulting from acute CO_2 retention in asthma indicates a life-threatening emergency. With the same pH in diabetic coma, most patients will recover without therapy (e.g. HCO_3^-) specifically directed to correcting the acidemia.

mination to prove that the nephritic patient has accumulated fixed acid; nor, in the case of the vomiting ulcer patient whose plasma $[HCO_3^-]$ is 35, is proof required that HCl has been lost from the body. Simple clinical observation immediately determines what is the primary acid-base disturbance.

But what if the nephritic has been vomiting? What if the ulcer patient has been taking $NaHCO_3$? It is true that pH in both these patients will tend to be returned toward normal. But who would use the pH to determine the primary versus the secondary or complicating effects? Again clinical observation is what is needed, not a pH measurement.

Measurement of pH is needed to determine the ratio $[HCO_3^-]/Pco_2$ and it is this ratio, together with the value of both numerator and denominator, which aids in the understanding of acid-base disturbances not so obvious as in the clinical example just given. Robinson's[101] parody is particularly useful here:

$$pH = pK + \log \frac{\text{KIDNEYS}}{\text{LUNGS}}.$$

Kidney failure (which primarily lowers the numerator) is made up for by secondary (compensatory) hyperventilation which lowers the denominator by lowering Pco_2. Lung failure (which primarily increases the denominator) is made up for by secondary renal retention of HCO_3^-.

The direction of these primary disturbances and secondary responses of the lungs and kidneys is summarized in Figure 20. The pH at a particular time during an illness (especially if repeated at another time) together with an understanding of the time course of secondary responses can be, as will be shown in the clinical chapters to come, of considerable value in determining the primary cause of the acid-base disturbance under investigation. However, no single index (including the "base excess") can be used intelligently in isolation. The other acid-base variables and, even more important, other *clinical* information are necessary to distinguish primary from secondary events.

A recent clinical example shows the use of a pH. A 27-year-old male was recovering from a cerebral contusion following a motorcycle accident. The neurosurgeon had requested that the Pco_2 be kept above normal to avoid alkalemia and alkalosis of the brain and the respiratory care service had complied with this request. The patient was still only semiconscious three weeks after but had begun to breathe adequately on his own and had been removed from the ventilator. A sudden fever of 41°C occurred (as a result of a candida infection occurring consequent to antibiotic therapy) and he became tachypneic and appeared to be hyperventilating.

The neurosurgeon feared "profound sepsis" and ⟨metabolic acidosis⟩. A CO_2 content (venous) was 16.5 mEq/L and an order was written for $NaHCO_3$, one ampule (44 mEq/) q 2 h × 6. The respiratory care service was called to evaluate the ⟨acidosis⟩.

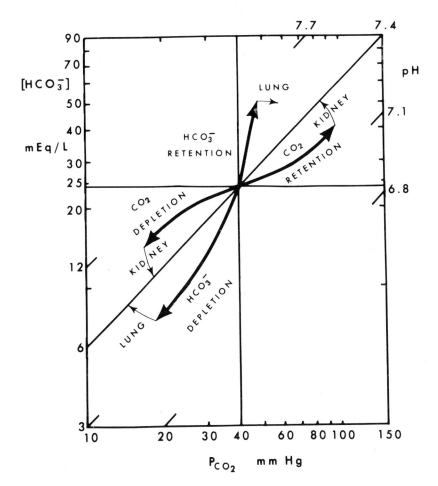

FIG. 20. Summary of primary (heavy arrows) acid-base disturbances and secondary responses (light arrows). The arrows show the general direction of the changes with time and not the exact paths followed. Only pure (unmixed) primary disturbances are depicted in this figure. As will be shown in Chapter 5, the upper right and lower left quadrants can, in mixed disturbances, be occupied because of other than secondary or compensatory processes. Mixed disturbances, which together produce the greatest pH deviations, tend to occupy the upper left and lower right quadrants. (See Fig. 41.)

By the time an arterial blood sample was drawn, 44 mEq $NaHCO_3$ had been given (by intravenous "push"). The patient continued to hyperventilate. The arterial blood gas analysis showed:

pH	7.50	$[HCO_3^-]$	19.4 mEq/L
P_{CO_2}	25 mm Hg	BE	−2
P_{O_2}	77 mm Hg		

These results showed that the venous CO_2 content was low not because of ⟨acidosis⟩ but because of *primary CO_2 depletion*. The hyperventilation had so lowered the P_{CO_2} that the pH rose to 7.50.

Was it necessary to obtain an arterial blood sample to diagnose respiratory alkalosis? No, because a free-flowing (non-congested) venous sample might also have had a high pH and a low P_{CO_2}; its BE would have been near zero. However, the sicker the patient (and this was a sick patient) the less informative is a venous sample for acid-base measurement.

Without a pH measurement (preferably arterial) this patient might well have been quite inappropriately treated.

Isolated measurements such as BE, $[HCO_3^-]$ or even pH are seldom adequate when obtained singly or considered by themselves. As Clark wrote: "A measurement of but *one* quantity found in any *one* of the following equations cannot define the state of the carbonic acid system.

$$pH = pK_1' + \log \frac{[HCO_3^-]}{Ko \; P_{CO_2}}$$

$$pH = pK_1' + \log \frac{[Total \, CO_2] - Ko \quad P_{CO_2}}{Ko \; P_{CO2}}$$

Remember this, oh ye of little faith, when, in a puzzling case, you are asked to judge the state of affairs from *one* datum."[102]

4

··

Oxygenation and the Arterial Po₂

"Is the quantity of oxygen taken up by the cell conditioned primarily by the needs of the cell, or by the supply of oxygen? The answer [is] clear, the cell takes what it needs and leaves the rest. Respiration therefore should be considered in the following sequence. Firstly the call for oxygen, secondly the mechanism by which the call elicits a response, the immediate response consisting in the carriage of oxygen to the tissues by the blood and its transference from the blood to the cell. Thirdly in the background you have the further mechanism by which the blood acquires its oxygen. It is not the habit of writers on respiration to adopt this order, quite the contrary, but their reason for placing pulmonary respiration in the foreground of the picture is a purely practical one—pulmonary respiration is more evident, both to the eye and to the understanding. I imagine they will not quarrel with me if I make an attempt to treat the matter in what appears to be its logical sequence and make some estimate of the call which the blood has to meet, before entering into a discussion of how the call is to be met." JOSEPH BARCROFT (1914)[103]

··

ALTHOUGH the partial pressure of oxygen of arterial blood (PaO₂) is becoming the most widely used measure of hypoxemia, arterial oxygenation represents only the first stage in the transport of oxygen from the blood to the mitochondria of vital organs. It is true that arterial Po₂, being mainly determined by the manner in which the lungs are ventilated and perfused, is a useful index for diagnosing respiratory failure. But the oxygenation of the tissues served by the lungs is only indirectly influenced by PaO₂. This is because

75

the ratio of O_2 supply to O_2 demand and the P_{O_2} of these tissues may be normal even though the arterial P_{O_2} is low. There may, in other words, be *arterial hypoxemia* without *tissue hypoxia*. The arterial P_{O_2} is only one index of oxygenation measured at only one site in the body.

It will be found, however, that when the significance of oxygen partial pressures in the body is understood, the arterial P_{O_2}, *when considered along with other blood gas measurements and clinical signs* can be of great importance in the diagnosis and therapy of a wide variety of clinical conditions.

In this chapter we will define four measurable indices of oxygenation, indicate the nature of oxygen demands and discuss the normal means of oxygen transport and their adjustments in disease.

INDICES OF OXYGENATION AND OXYGEN DEMANDS

Four Measurable Indices of Oxygenation

Molecular oxygen, by which the energy of foodstuffs is liberated, exists in body fluids either free (as dissolved O_2) or loosely combined with respiratory pigments like hemoglobin. Oxygen is extremely insoluble in most body fluids. Hence the free O_2 is in very low concentration even in arteries, where it is at its maximum in blood when air is being breathed. It is best considered as an intensity factor and expressed as the partial pressure of oxygen, P_{O_2}, as in Figure 21.

The concept of P_{O_2} as a potential in body tissues is a dynamic one. As blood flows through a tissue, oxygen tensions can vary greatly along a capillary—from a P_{O_2} at the arterial end of about 100 to a venous P_{O_2} of about 40 mm Hg. The driving or unloading potential (the gradient between blood and tissue) clearly falls off markedly toward the venous end of the capillary.

The quantity of oxygen needed for metabolism is made available by being transported in blood as oxyhemoglobin, O_2Hb. The process of oxygenation is illuminated* by fully understanding the distinction between the quantity of

---

* Similar illumination was brought to the process of heat transport by a practicing physician (Joseph Black) 200 years ago when he showed that the distribution of heat quantity among different bodies was non-uniform even though the bodies were at the same temperature.

"The nature of this equilibrium was not well understood until I pointed out a method of investigating it. Dr. Boerhaave imagined that when it obtains, there is an equal quantity of heat in every volume of space, however filled up with different bodies. . . .

"But this is taking a very hasty view of the subject. It is confounding the quantity of heat in different bodies with its intensity (temperature), though it is plain that these are two different things, and should always be distinguished when we are thinking of the distribution of heat."[104]

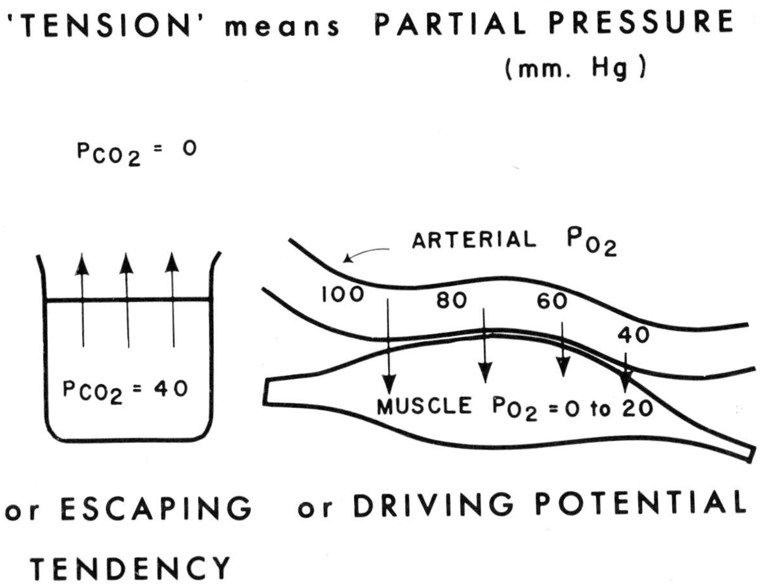

'TENSION' means PARTIAL PRESSURE
(mm. Hg)

$P_{CO_2} = 0$

ARTERIAL P_{O_2}

100   80   60   40

$P_{CO_2} = 40$

MUSCLE $P_{O_2} = 0$ to 20

or ESCAPING   or DRIVING POTENTIAL
TENDENCY

FIG. 21. Physically dissolved respiratory gases have the character of intensities more than of quantities. The use of the word "tension" (to stand for partial pressure) is almost restricted to pulmonary physiologists. It probably reflects the need to emphasize the concept of physically dissolved gases as intensities in contrast to O_2 and CO_2 quantities such as O_2Hb and $HCO_3{}^-$.

In between capillaries, P_{O_2} falls markedly. The P_{O_2} midway between two cat brain arterioles 100 μ apart is much lower than that adjacent to them.[105]

oxygen in different parts of the body and the intensity of oxygen (its P_{O_2})— a distinction similar to that we have made for the case of carbon dioxide, but even more striking.

We have already contrasted carbon dioxide as a quantity of material (mostly as bicarbonate) and the P_{CO_2} as an escaping tendency dependent upon physically dissolved molecular CO_2. In the present discussion the contrast to be emphasized is that HCO_3^- is unevenly distributed in the body and molecular CO_2 almost uniformly distributed. (See Fig. 22A.)

In the case of oxygen, O_2Hb is very unevenly distributed in the body. As a quantity of material required for energy production, oxygen exists almost exclusively in oxyhemoglobin or oxymyoglobin. Dissolved molecular O_2 is in even lower concentration than dissolved CO_2. Finally, as shown in Figure 22B, large *P_{O_2} tension gradients* may exist. In fact these must exist for adequate oxygen transport. Such tension gradients are best thought of as driving potentials, a term similar to the "diffusion pressures" of Haldane.[1]

From the foregoing it is clear that the transport of oxygen to the cells of a

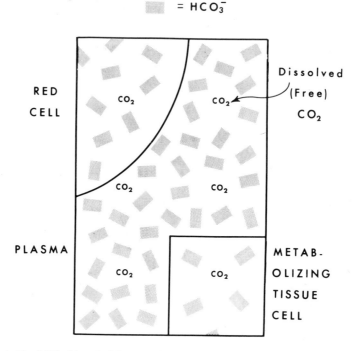

 = HCO_3^-

RED
CELL

CO_2 CO_2

CO_2 CO_2

PLASMA

CO_2 CO_2

Dissolved
(Free)
CO_2

METAB-
OLIZING
TISSUE
CELL

FIG. 22. A. The [HCO_3^-] in arterial plasma is normally 24 mEq/L and about 14 mEq/L in red blood cells. It is probably lower than 14 in most tissue cells. Molecular CO_2 is nearly uniformly distributed in the body as indicated in the figure and Pco_2 gradients are small.
For every mm Hg of Pco_2, only 0.03 mEq of CO_2 is dissolved in a liter of plasma.

tissue depends on the *local* O_2 driving potential. Physicochemical and physiologic devices of the greatest ingenuity provide for the maintenance of this local Po_2. The challenge to understand these devices has been accurately stated by Kety:

> "The maintenance of an optimal oxygen tension *about each cell* is one of the cardinal functions of the circulatory and respiratory systems of higher animals, and an adequate explanation of the mechanisms involved in its accomplishment constitutes a large segment of physiological knowledge."[106] (Italics mine.)

The two usually employed indices of oxygenation, % saturation and Po_2, are best displayed by the classical oxyhemoglobin dissociation curve and their characteristics illustrated by two fairly straightforward clinical examples—anemia and carbon monoxide poisoning.

In Figure 23 the usual Po_2-% saturation plot is shown for normal human blood. One point on this curve represents the (mixed) venous blood % saturation and its P_{O_2}. At this point the two indices reflect, in different ways, the state of oxygenation of the "body as a whole." Before discussing the use of such indices, it is necessary to define them.

Index 1. Percentage oxygen saturation

$$\% \text{ Sat} = 100 \times \frac{\text{Quantity of } O_2 \text{ bound to the hemoglobin of a blood sample}}{\text{Quantity of } O_2 \text{ that can be bound by this hemoglobin}}.$$

This index is thus the ratio of oxygen content (minus physically dissolved O_2) to oxygen capacity (minus physically dissolved O_2). It is best to think of this index as it is classically determined, that is, by Van Slyke extraction of gas from anaerobically drawn blood and measurement of the O_2 content followed by artificial exposure of an aliquot of the blood to air or oxygen and a second Van Slyke extraction to measure O_2 capacity. Clinically a colorimetric estimate of these time-consuming determinations is suitable for most purposes. In the absence of carboxy-hemoglobin or other abnormal hemoglobin the

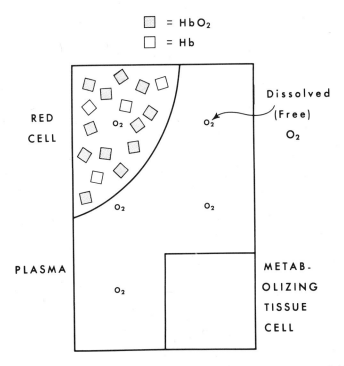

FIG. 22. B. Oxygen in high concentration exists only in red cells (or oxymyoglobin). Very little oxygen is dissolved in plasma. The P_{O_2} inside cells is very low. A clear statement of the concept of tissue P_{O_2} has been given by Comroe.[107]

For every mm Hg of P_{O_2}, only 0.03 cc of oxygen is dissolved in a liter of plasma.

NORMAL O₂ INTENSITY— QUANTITY RELATIONSHIP

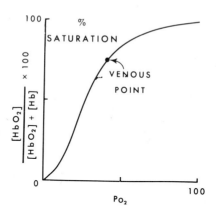

FIG. 23. Mixed venous blood normally is 75% saturated with O_2 and has a Po_2 of 40 mm Hg. As discussed in the text, % saturation is not strictly a concentration but a ratio of concentrations. With normal hemoglobin concentration, % saturation can represent O_2 quantity or the ordinate as contrasted to O_2 intensity in the abscissa. (With permission.[164])

following formula is equivalent to the previous one:

$$\% \text{ Sat} = 100 \times \frac{[O_2Hb]}{[O_2Hb] + [Hb]}$$

where [Hb] is the concentration of reduced hemoglobin (i.e. able to bind O_2).
   Index 2. Oxygen partial pressure, mm Hg

$$Po_2 = \frac{1}{.03} \times \text{Physically dissolved } O_2 \text{ (in cc/L)}*$$

The pressure which oxygen exerts in blood depends on the concentration of O_2 physically dissolved in plasma. Blood Po_2 does not depend simply on the concentration of O_2Hb nor does it vary linearly with % saturation. It may not even change when the total quantity of O_2 in the blood changes. To explain how Po_2 can vary independently of both O_2 saturation and of the quantity of oxygen in the blood, it is necessary to analyze the meaning of % saturation.

* Oxygen is much less soluble than CO_2 in plasma. Solubility factors for O_2 in plasma and in whole blood vary with temperature.[4,108] For plasma at 38°C, a factor of 0.003 can be used when free or dissolved O_2 is expressed as "vol %." The solubility factor of 0.03 is used here to express the fact that 0.03 cc of O_2 are dissolved per liter of plasma for every mm Hg Po_2. The numerical equivalence of this solubility factor with that for CO_2 is convenient. Thus $0.03 \times Pco_2 = [\text{Free } CO_2]$ in mEq/L; $0.03 \times Po_2 = [\text{Free } O_2]$ in cc/L. A complete discussion of gas solubility coefficients has recently been published.[109]

It is clear that % saturation, being a ratio of concentrations, cannot be a measure of concentration. In other words, a low saturation, even though this is called "hypoxemia," need not mean that the oxygen concentration in blood is low. For example, a polycythemic sample (hemoglobin 22 g) has an O_2 capacity of 300 cc/L. This may be only 60% saturated and contain as much O_2 (i.e. 180 cc/L) as a normal sample (hemoglobin 15 g) with an O_2 capacity of 200 cc/L which is 90% saturated. On the other hand, a normal saturation may be associated with a low O_2 content. In anemia, for example, % saturation may be normal, yet the quantity of O_2 per unit volume of blood is low. Changing the scale on the vertical axis to cc of O_2 per liter of blood shows, as in Figure 24, how anemia affects O_2 quantity. Oxygen intensity (Po_2), however, is not directly changed by the dilutional effect of anemia.

A complete contrast to anemia is shown in Figure 25. The venous point of normal blood (closed circle) is seen to shift to the left at a given saturation of oxygen when CO has been added to blood. This shift obviously interferes with the ability of the blood to release its oxygen to the tissues. This effect of CO is different from the "anemic effect" usually thought of in CO poisoning. It is true that CO ties up hemoglobin as COHb and prevents it from carrying oxygen. But the effect on the shape of the O_2 dissociation curve is different and perhaps more important clinically.

EFFECT OF ANEMIA
ON O₂ QUANTITY

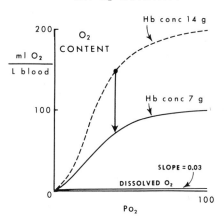

FIG. 24. Anemia always lowers the O_2 concentration. This lowering would not be seen if the ordinate were expressed in the usual units of % O_2 saturation as in Figure 23. Anemia, *per se*, does not affect the Po_2 of blood. Secondary changes, particularly in cardiac output, may change venous Po_2 by affecting the arteriovenous O_2 concentration difference. Indeed, if cardiac output did not rise in severe anemia, severe lowering of venous O_2 concentrations and tensions would occur (see Fig. 29). (With permission.[164])

EFFECT OF CO
ON O_2 INTENSITY

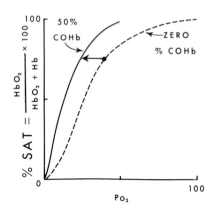

FIG. 25. Carbon monoxide somehow makes the hemoglobin molecule bind its O_2 more tightly. Hence the O_2 unloading potential of O_2Hb is reduced. The horizontal arrow shows how the % saturation with oxygen is not changed by CO poisoning when HbO_2 and total hemoglobin capable of binding O_2 are reduced proportionally. (With permission.[164])

Thus CO poisoning imperils tissue oxygenation by a kind of double jeopardy as shown by Figure 26. Not only does it prevent adequate amounts of O_2 from being delivered to the capillaries per liter of blood, but the O_2 which arrives at a tissue capillary *cannot be properly released* by the hemoglobin—a circumstance reflected by a lowered Po_2. Because the (capillary and venous) blood Po_2 in CO poisoning is low for a special reason, this type of hypoxemia could be designated by a special name (see reference to Barcroft in footnote, p. 83); we will simply point out now that a low Po_2 should be thought of as having a different significance than % saturation even though both are called hypoxemia.

We are now in a position to define a third index of oxygenation—the quantity of oxygen per liter of blood. Unfortunately an absolute determination of this quantity is a time-consuming procedure requiring extraction and measurement of oxygen gas.[56] For practical purposes we therefore base this third index on measurements that can be made colorimetrically (the hemoglobin concentration and optically determined % saturation).

Index 3. Quantity of O_2 in blood, cc/L

$$O_2 \text{ content} = 1.34 \times [Hb] \times \% \text{ Saturation}/100$$

where the factor 1.34 represents the number of cc of O_2 which 1 g of normal hemoglobin can carry when fully saturated, and [Hb] the hemoglobin concentration in g/liter. Concentrations of O_2 and hemoglobin in blood are often

EFFECT OF CO
ON BOTH

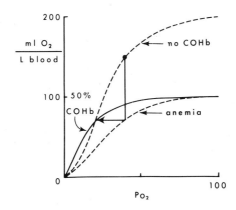

FIG. 26. The solid curve is in the actual position of the O_2 dissociation curve in a case of CO poisoning on appropriate intensity-quantity coordinates.[102] The upper dotted curve is the curve for blood without any CO in it. The lower dotted curve is that of anemic blood (or that of blood partly saturated with CO were there no CO effect on hemoglobin's ability to release O_2). (With permission.[164])

expressed per 100 ml blood. Expressing them per liter is in line with recent recommendations.[26] It also facilitates calculation of O_2 delivery since $cc\text{-}L^{-1} \times L\text{-}min^{-1} = cc\text{-}min^{-1}$.

The fourth index requires a knowledge of the rate of blood flow in liters per minute either to the entire body (cardiac output) or to a particular tissue.

Index 4. Oxygen delivery, cc/min

$$O_2 \text{ delivery} = O_2 \text{ content} \times \text{Blood flow}$$

The concept of the rate of oxygen delivery was introduced into clinical medicine by Richards.[110] It is a basic physiologic variable which has important clinical implications as well as reflecting the third of the three classical types of hypoxemia outlined by Barcroft,[111] i.e. anoxic, anemic and stagnant. A fourth type was also suggested by Barcroft even earlier as being associated with a lowered ability of hemoglobin to unload its oxygen.* Following Bar-

---

* In 1914 Barcroft[103] coined the word "pleonectic" to refer to a left-shifted O_2 dissociation curve, i.e. one reflecting a lowered ability of hemoglobin to release its oxygen (high O_2 affinity). Pleonexia is from πλεονεξια, a disposition to take more than one's share; to be grasping. We will include this type in the term "potentiometric hypoxemia," recognizing that blood Po₂ can be low not only because it is equilibrated with a low Po₂ (e.g. arterial blood at high altitude), but also because the hemoglobin is so "stingy" that if any O_2 leaves the blood the Po₂ falls precipitously (i.e. in venous blood in CO poisoning).

"Histotoxic anoxia," e.g. from cyanide poisoning, was a special term coined by Peters and Van Slyke.[28]

croft's lead, we associate changes in our four indices of oxygenation with four types of disturbed oxygen transport as shown in Table 5.

TABLE 5. *Abnormalities of oxygenation*

Index	Type	Clinical Signs
1. Low O_2 delivery (cc/min)	Ischemic	Coolness, pallor or cyanosis
2. Low O_2 content (cc/L)	Anemic	Pallor
3. Low arterial saturation (%)	Colorimetric	Cyanosis
4. Low arterial Po_2 (mm Hg)	Potentiometric	None specific

The order is chosen to emphasize the primary importance of tissue oxygenation. The first two abnormalities usually cause hypoxia of peripheral tissues; the last two are types of "central" or "hypoxic" hypoxemia.

1. When blood flows so slowly that the rate of O_2 delivery to a tissue is lower* than the rate of O_2 consumption by the tissue, O_2 is extracted in abnormally large quantity from the capillaries. Venous blood becomes hypoxemic and the tissue hypoxic because of low O_2 delivery.

2. In the anemic type "the quantity of functional hemoglobin is too small."[111] As suggested by Barcroft the type includes hypoxemia caused by CO poisoning which lowers the quantity of O_2-carrying hemoglobin and hence the O_2 content.

3. A lowered saturation can be determined gasometrically (as the ratio of O_2 content to O_2 capacity) as well as colorimetrically but "colorimetric" is chosen to characterize this index to stress the difference between O_2 content and saturation.

4. A lowered Po_2 is usually associated with a low saturation, but the relation of the two is not constant, being mainly affected by pH and the intracellular composition of erythrocytes. The Po_2 is measured potentiometrically (being a pressure or intensity factor).

* If it *stays* lower, anoxia would result; what usually occurs is a new steady state in which O_2 consumption proceeds at a normal rate but at a low tissue Po_2.

Oxygen Demands

Molecular oxygen is the sink toward which electrons flow along the respiratory chain in mitochondria and it must accept these electrons at the rate at which they leave the chain to maintain the rate of aerobic energy production needed for life. Oxygen demands, therefore, should be expressed as rate of O_2 uptake. In certain circumstances, illustrated by Figure 37, this rate of O_2 uptake may conveniently be expressed as a percentage of the maximum rate which a tissue or an animal is capable of achieving.

Oxygen consumption, as stated in the quotation at the beginning of this

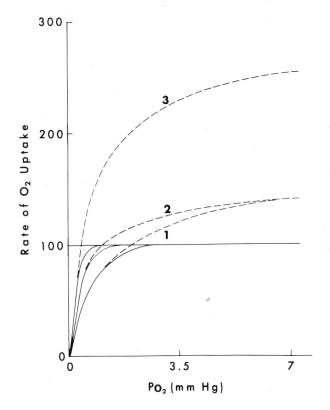

FIG. 27. Cellular O_2 uptake is scarcely affected until Po_2 becomes very low. Curve 1: tumor cells[112]; curve 2: liver cells[113]; curve 3: isolated liver mitochondria.[114] (Adapted from Jöbsis.[115])

chapter, is "conditioned primarily by the needs of the cell" and these needs are met even when the Po_2 of the cell is extremely low as shown in Figure 27. Except in relatively isolated tissues (e.g. in a Warburg apparatus) or under special conditions for measuring respiratory gas exchange as in a BMR determination, oxygen demands must be estimated indirectly. For physiologic and clinical work, the ratio of oxygen supply to demand (delivery to consumption) as expressed by venous Po_2 has been found to be a useful index:

$$PvO_2 \propto \frac{O_2 \text{ delivery}}{O_2 \text{ consumption}}$$

Like other measures of intensity, the venous Po_2 is proportional to a ratio of quantities, i.e. to the reciprocal of the O_2 utilization fraction. (The above formula can be written $PvO_2 \propto \dfrac{CaO_2 \times \dot{Q}}{(CaO_2 - CvO_2) \times \dot{Q}}$ or as $PvO_2 \propto$

$CvO_2 = CaO_2 - \dot{V}o_2/\dot{Q}$ where $CaO_2 - CvO_2$ is the arteriovenous O_2 concentration difference and $\dot{Q}$ is the blood flow.)

When the *mixed* venous blood Po_2, $P\bar{v}O_2$, is 40 mm Hg or above (about 75% saturated or more) the oxygenation of the body is usually considered to be normal and a $P\bar{v}O_2$ less than 30 to indicate tissue hypoxia. The PvO_2 of the various tissues varies greatly, being directly proportional to the ratio of O_2 supply to O_2 demand as shown in Table 7. The demands of several tissues will now be considered.

Oxygen demands in cells and tissues.   The Po_2 of venous blood draining a tissue gives an indication of interstitial oxygen tension, but inside cells the Po_2 is much lower; inside mitochondria it is lower still, probably near 1 mm Hg.[116] "There are unquestionably oxygen tension values below which cell function is compromised,"[117] but what these values are *in vivo* is unknown. As indicated in Figure 21, Po_2 tension *gradients* must exist in and around cells. Hence to state a single figure for optimal cellular Po_2 has been an elusive goal. This is evident from the use of the term "critical Po_2," namely, a figure (about 3.5 mm Hg)[115] below which the O_2 uptake begins to fall.[118]

Ignorance of the exact intracellular Po_2 needed for normal cell life should not blind us to considerations which are probably of even greater importance, namely, adaptation, or adjustment, of cellular and tissue demands to variations in oxygen supply. Such adjustment may occur by intracellular mechanisms and by virtue of tissue organization. Individual cells can develop astonishing adaptations to adverse conditions of obvious relevance to medicine, either by making better use of the energy available or by decreasing their energy demands.[119] The biochemical machinery of individual cellular adaptation remains obscure despite the hopeful predictions of leaders in the field.[120] Yet the capacities of organized tissue to adjust to variations in the oxygen supply are at the basis of mechanisms of immediate clinical importance. Thus myocardial oxygen requirements are determined by measurable hemodynamic factors which can be altered by nitrites, β-adrenergic blockade and carotid sinus nerve stimulation[121] and the success of propranolol treatment of the stagnant (ischemic) hypoxia of coronary artery disease is explained by reducing myocardial oxygen requirements caused by exertion.[122-126]

In organ systems.   Although the body's demand for O_2 is the fairly constant sum of the separate demands of each organ, these separate demands can vary enormously between organs and in respect to time. Table 6 shows that the brain, only 2% of the body weight, consumes 20% of the body's oxygen uptake. This O_2 demand is met by a brain blood flow of only 14% of the cardiac output.[127] "The brain shares with the heart the dubious distinction of having a metabolic requirement disproportionate to its blood supply."[128] In addition, central nervous tissue is inherently sensitive to lack of sufficient oxygen, surviving only a few minutes of anoxia.

TABLE 6. *The O_2 demands of three organs in man at "rest"*[117,128,129]

	O_2 Consumption (% of total)	Revivable Post Anoxia (minutes)
Brain	20	5
Heart	9	10
Kidney	6	60

The O_2 consumption $= \dot{Q} \times (CaO_2 - CvO_2)$ where $\dot{Q}$ is blood flow and $CaO_2 - CvO_2$ is the arteriovenous O_2 concentration difference.

How are the oxygenation indices we have defined related to the demands for oxygen by these three organs? In Table 7, it is seen that the Po_2 is lowest in the coronary veins, intermediate in the jugular veins and highest in the renal veins. These venous Po_2 values are generally taken to reflect fairly closely the "tissue oxygen tension" of these organs. It should be noted that this tension correlates with the oxygen delivery index defined above. Assuming a normal arterial O_2 content, the ratio of the rate of oxygen delivery to the oxygen consumption of the organ is clearly related to the effluent venous Po_2. (The reciprocal of this ratio is sometimes called the O_2 utilization fraction.)

In the body-as-a-whole. The overall O_2 requirements of the body are, in some ways, easy to describe. A resting 70 Kg man uses 250 cc O_2 per minute. This normal metabolic rate must be increased 10-fold for moderately strenuous exercise, either during the exercise or in recovery when the O_2 debt is paid. Common clinical conditions (thyrotoxicosis, fever, the extra work of breathing associated with trauma to the chest or abdomen) also clearly increase the demands for oxygen.

TABLE 7. *The "oxygen environment" of three organs as reflected by their venous Po_2*

	Effluent Venous Po_2 (mm Hg)	O_2 Delivery (cc/min) / O_2 Consumption (cc/min)
Heart	18	0.8
Brain	35	2.8
Kidney	62	14.0

The "oxygen environment" of three organs as reflected by their venous Po_2 is a function of the ratio of O_2 delivery to O_2 consumption. This ratio is $\dot{Q} \times CaO_2/\dot{Q} \times (CaO_2 - CvO_2)$, i.e. the reciprocal of the O_2 utilization fraction.

But in other clinically important circumstances the metabolic rate of the body is determined by poorly understood factors. In states of starvation, sleep, anesthesia and hypothermia, for example, the need for O_2 is often (but not always) depressed[130-134]; under special conditions of stress,[135,136] obscure mechanisms for reducing the body demands for O_2 seem to operate; for example, animals adapted to trauma demand less O_2 in response to more trauma than do control animals[137] and O_2 demands can be reduced by as much as 15 to 25% for many hours after acute experimental CO_2 retention.[138] Under conditions of controlled hyperventilation the total O_2 uptake of man is reduced by breathing CO_2, and increased by hypocapnia.[139] Even in quite ordinary circumstances (as in patients with airway obstruction simply walking naturally), extra O_2 demands imposed by disease are satisfied by unexplained economies. Thus, during muscular exercise, emphysematous patients require no more oxygen than do normal persons.[140] Even though the O_2 cost of their breathing is greatly increased, such patients may or may not accumulate an increased O_2 debt.[141]

In summary, O_2 demands not only vary between organs but the total bodily demands can change under various conditions. Some of these conditions are the result of disease; others, as will be shown in Chapters 5 and 6, can be imposed, knowingly, or unknowingly by modern therapy. The oxygen environment of tissues, reflected by PvO_2, will depend on how adequately O_2 demands are met by the O_2 supply, i.e. by O_2 transport processes. During stress or illness these processes act as reserve mechanisms.

OXYGEN TRANSPORT RESERVE MECHANISMS

The four indices of Table 5 can be used to categorize four major O_2 transport mechanisms and to show not only how they are impaired in disease but also how each can act as *reserves*, i.e. can compensate (acting alone or in combination) for impairment in a particular mechanism. These reserves are (1) circulatory, (2) erythropoietic, (3) chemical and (4) pulmonary. (See Fig. 28.) They will be taken up in this order (from the least to the best understood) because many of the most important oxygenating mechanisms are obscure as to mode of action, difficult to quantitate, and hence perhaps should be thought of first.

1. *The Circulatory Reserve*

a. *The microcirculation; capillarity.*   The anatomic fact that few if any living mammalian cells are located more than 0.05 mm from a blood capillary attests

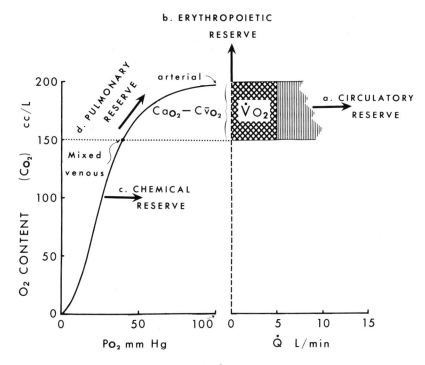

FIG. 28. The oxygen demands of resting man, $\dot{V}_{O_2}$, are normally satisfied by a cardiac output ($\dot{Q}$) of 5 L/min at an a-vO_2 diff ($CaO_2 - C\bar{v}O_2$) of 50 cc/L. Thus $\dot{V}_{O_2} = \dot{Q}(CaO_2 - C\bar{v}O_2) = 5 \times 50 = 250$ cc/min. The product of the flow and the a-v difference is illustrated by the cross-hatched area on the figure. (Adapted from Metcalfe.[142])

a. Use of the circulatory reserve *permits* $\dot{V}_{O_2}$ *to rise at the same a-v difference,* a powerful means of maintaining a high $C\bar{v}O_2$ and tissue Po_2 in the face of increasing O_2 demands. Only the total flow is depicted; clearly, local increases in flow would have the same local effect.

b. Increasing the hemoglobin increases *the capacity of blood to carry* O_2 *at a given Po_2.* Erythropoiesis can hence increase the scale of the ordinate, raising tissue Po_2, other factors remaining the same.

c. Chemical changes (low pH, high intracellular phosphate) in red cells can increase *the potential for O_2 unloading* at any saturation. This unloading potential is expressed by the Po_2 at 50% saturation, i.e. the P_{50}.

d. The pulmonary reserve is hyperventilation. Normal lungs can *increase the arterial Po_2* (and usually the CaO_2). They provide the major means of "staying high on the dissociation curve."

to the unique dependence of oxygen transport on the proximity of oxygen-containing blood to oxygen-demanding cells. Adequate description of oxygen diffusion from blood to tissues requires knowledge of "the three dimensional distribution of the minute vessels and of the various cell types in tissues. There is an extraordinary paucity of information of both distributions . . . "[143]

Estimates of capillary density even in the best studied tissue (muscle) vary greatly and are usually expressed as the number of capillaries per *square* millimeter (resting muscle 200 to 400, exercising muscle 600 to 5000 per mm²).[144]

That an increase in capillary density is a means of compensating for arterial hypoxemia is evidenced by the fact that exposure to high altitudes increases this density in muscle.[145] Its importance was dramatized by the remarkable finding of Stainsby and Otis[146] that the critical arterial Po_2 was lower in exercising than in resting muscle; in other words, arterial hypoxemia had to be more severe to lower O_2 uptake by exercising than by resting muscle. How the effective capillary bed widens in response to hypoxia and exercise is still best described in the words of the last-named authors: "Since the response occurs in the absence of extrinsic innervation, it must be autonomous, or in present terminology, autoregulated. The mechanism is unknown."

b. *Regional blood flow.* The ability of each organ to satisfy its oxygen demands is almost automatic (autonomous) by virtue of the fact that arterial hypoxemia causes immediate dilatation of systemic arterioles to that organ. At a given systemic pressure, such local dilatation favors preferential local perfusion. When man breathes 10 to 13% oxygen, the brain increases its blood flow 35%[147]; coronary vessels are even more responsive to low arterial Po_2, but the intestinal renal and cutaneous vessels far less so.[148]

The physician should be aware of the power of *coordinated* central and local vasomotor responses to blood gas changes in general. Thus cerebral vessels dilate even more per mm Hg increase in arterial Pco_2 than with equivalent reductions in arterial oxygen tension; an increased Pco_2 (which very often accompanies low Po_2 in the blood) is probably the most effective means of preventing cerebral hypoxia during air-breathing because it not only dilates cerebral vessels but also can greatly increase systemic arterial blood pressure. Interference with blood flow to the vasomotor center in the brain produces a dramatic "CNS ischemic response" by sympathetic vasoconstriction of almost all systemic vessels. "The degree of sympathetic vasoconstriction caused by intense cerebral ischemia is often so great that some of the peripheral vessels become totally or almost totally occluded. The kidneys, for instance, will entirely cease their production of urine because of arteriolar constriction in response to sympathetic discharge."[127] "Sympathetic control is not generalized but appears to be designed for highly differentiated action" is an important conclusion strongly supported by recent work in animals[149]; local hypoxia and tissue acidity undoubtedly modulate generalized sympathetic action by overriding vasoconstrictor stimuli in areas where oxygen demands are especially high.

Sudden augmentations of local blood flow in response to low arterial Po_2

can "steal" blood from a less favored tissue. In such a tissue, therefore, ischemic hypoxia may be superimposed on arterial hypoxemia unless more blood can be supplied to all the tissues. This brings us to the major means of calling on the circulatory reserve.

c. *Cardiac output.* Increasing the cardiac output as a reserve mechanism in the defense against tissue hypoxia is a relatively expensive compensation. In resting man the heart uses 9% of the O_2 available to the body; the heart's $\dot{V}_{O_2}$ may rise 8-fold and consume more than 50% of the total. Of the heart as a means of preventing venous P_{O_2} from falling below 30 mm Hg, "a tolerable value for long term survival," Tenney and Lamb write, "If the energy required for moving a quantity of air into the lungs is roughly equal to the energy necessary to move an equal volume of blood out of the right heart (which is probably an overestimation of the relative work of breathing) and if the left heart work is four times the right heart work, then there would be a predicted fivefold difference in the energy required to accomplish the same $P\bar{v}_{O_2}$ during moderate hypoxic hypoxia by means of the circulatory as compared with the ventilatory route."[117]

Because of the nature of this circulatory reserve, it tends to be called on more in ischemia and anemia than in arterial hypoxemia. This is probably because the first two are more often associated than the latter with a low *venous O_2 content* which the cardiac output largely controls. These relationships are shown in Figures 28 and 29. In anemia, for example, increasing the cardiac output, $\dot{Q}$, ($\dot{V}_{O_2}$ constant and CaO_2 constantly low), tends to raise mixed venous O_2 content, $C\bar{v}O_2$, in accordance with the Fick equation:

$$\uparrow\dot{Q} = \frac{\dot{V}_{O_2}}{CaO_2 - \uparrow C\bar{v}O_2}$$

Acutely induced anemia (hemoglobin falling to 5 g/100 ml) causes the cardiac output to increase after a few hours by about 30%,[150] the magnitude of change depending mainly on whether or not the blood volume has been maintained or reduced; in children with severe chronic anemia (hemoglobin 3.3 g/100 ml) the cardiac index is increased by 70%.[151] By contrast the cardiac output response to high altitude hypoxemia is slight and transient, lasting less than a week.

If anemia is severe enough, cardiac output compensation is incomplete in the following sense: raising output alone does not prevent tissue hypoxia because the ratio of supply to and demand by the tissues, $CaO_2/(CaO_2 - C\bar{v}O_2)$, remains low and hence the venous O_2 tension is low. (This is usually referred to as stating that the coefficient of utilization rises in anemia.) However, in addition to the circulatory reserve, the chemical reserve (see below)

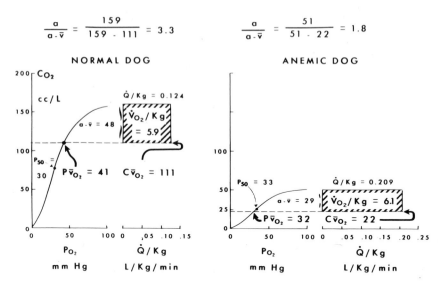

FIG. 29. The cardiac output increase to 0.209 L/Kg/min in chronically anemic dogs (50% of red cell mass removed) is not enough to deliver O_2 in proportion to the rate at which it is removed ($\dot{V}_{O_2}$). Therefore the supply-to-demand ratio $\dot{Q} \times CaO_2/\dot{Q} \times (CaO_2 - C\bar{v}O_2)$ falls from 3.3 to 1.8 (as the body is forced to increase the % O_2 utilization), and the $C\bar{v}O_2$ falls from 111 to 22 cc/L. Partly because the dissociation curve is shifted to the right (P_{50} rises from 30 to 33) the $P\bar{v}O_2$ in anemia is prevented from falling below 32 mm Hg. (From J. Metcalfe,[142] with permission.)

is called upon in anemia. As shown in Figure 29, circulatory and chemical mechanisms can combine to maintain the mixed venous Po_2 above 30 mm Hg.

2. *The Erythropoietic Reserve*

The realization that the red cell mass (in ml of RBC per Kg body weight) rather than the hematocrit is the proper measurement for assessing erythropoiesis[152] has clarified the old problem of whether the response to the hypoxemia of high altitude is different or not from the response of pulmonary patients with the same degree of hypoxemia; however, a new problem, regarding the mechanism of the erythropoietic response, has arisen. Most patients whose lung disease lowers their arterial O_2 saturation and Po_2 (and their O_2 content and delivery?) have red cell masses comparable (but not identical) to those of normal high altitude residents with similar arterial saturation.[153] As shown in the legend to Figure 30, the new problem is "which index of oxygenation is most closely related to the erythropoietic response"; recent work with hemoglobins which can[155] and cannot[156] easily release their oxygen strongly suggests that *tissue* oxygenation controls erythropoiesis.

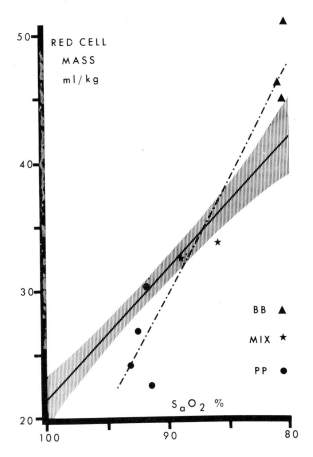

FIG. 30. The normal hematopoietic response in men residing at several altitudes (shaded area) and the response of patients with chronic airway obstruction (broken line). PP designates patients with near normal arterial saturation; BB patients have a low SaO_2%.[153]
The red cell mass measurements are linearly related to decreasing arterial O_2 saturation, (at least in this range of saturation) whereas their relation to arterial PO_2 is curvilinear, suggesting the possibility that hematopoiesis is a function of arterial O_2 content or O_2 delivery rather than of arterial Po_2. (Reproduced from the Journal of Clinical Investigation 47: 1627, 1968.)

In anemia, the mixed venous O_2 content ($C\bar{v}O_2$) is defended (prevented from falling precipitously) by increasing the cardiac output at a given hemoglobin level; at altitude, $C\bar{v}O_2$ is defended by increasing the hemoglobin level at a given cardiac output. Direct measurements in high altitude natives shown in Table 8 provide evidence that this is accomplished by maintaining a high O_2 delivery even during exercise at an altitude of 14,900 feet where both colori-

TABLE 8. *The maintenance of the O_2 delivery index at altitude*

		O_2 Saturation (%)	Hb (g/100 ml)	O_2 Content × (cc/L)	$\dot{Q}/m_2$ (L/min/m²)	= O_2 Delivery (cc/min/m²)	$C\bar{v}O_2$ (cc/L)
SEA LEVEL	Rest	95.7	14.8	190	3.97	756	150
	Exercise	94.9	15.4	197	6.83	1350	91
14,900 FEET	Rest	78.4	19.4	202	3.97	803	161
	Exercise	69.4	20.1	186	7.70	1430	83

The O_2 delivery index (sometimes called systemic oxygen transport) is maintained at or above sea level values by the polycythemia of altitude (avg. hematocrit 60%) as seen from the work of Banchero et al.[158], for this reason the $C\bar{v}O_2$ is maintained near or even above sea level values as shown in the last column.

metric and potentiometric hypoxemia exists even at rest, i.e. where arterial blood is 78.4% saturated and its Po_2 42 mm Hg.

During *acute* experimental hemorrhage or transfusion, the optimal hematocrit seems to be about 40%—optimal in terms of oxygen delivery, $CaO_2 \times \dot{Q}$. (Below 40% the output fails to compensate for acute anemia; above 40% the output falls.[157]) This does not seem to be true in chronic situations. Thus in long-term high altitude residents[158] and in polycythemic pulmonary patients,[154] even with hematocrits greater than 60%, cardiac output is maintained at normal levels. (Perhaps the *mechanism* is the action of hypoxemia in dilating peripheral arterioles and encouraging increased peripheral capillarity; perhaps the *price* is right ventricular hypertrophy resulting from hypoxic vasoconstriction of the pulmonary arterioles.)

3. The Chemical Reserve

Because it is the average Po_2 (not the arterial Po_2) in a systemic capillary which is responsible for driving oxygen to the tissues, it is clear that this average Po_2 could be raised to a higher level for a given arterial Po_2 if the venous Po_2 could be raised. (The average Po_2 can be estimated as $PvO_2 + \frac{1}{3}(PaO_2 - PvO_2)$,[159] or by the Bohr integration procedure.[160]) The rightward bulge in the lower part of the normal S-shaped oxygen dissociation curve of human blood almost certainly is an advantage in unloading oxygen under ordinary conditions.

Under the stress of exercise this lower bulge to the right increases. This bulge reflects an increased O_2 unloading potential (O_2 intensity) for a given saturation or O_2 content (O_2 quantity). The rightward shift occurs because the pH falls in plasma and (which is more important) inside red cells as increased CO_2 is pumped into the blood from working muscles. The shift is called the Bohr effect. (See Fig. 31.) (With severe exercise it is probable that local lactic acid production and increased temperature in the muscle further increase the O_2 unloading potential of blood.)

The current excitement over the fact that an increase in the concentration of a molecule, 2,3 diphosphoglycerate, (2,3 DPG),

$$
\begin{array}{l}
\text{COO}^- \\
| \\
\text{HC-O-}\textcircled{P} \qquad \textcircled{P} = \begin{array}{l}\text{phosphate}\\\text{residue}\end{array} \\
| \\
\text{H}_2\text{C-O-}\textcircled{P}
\end{array}
$$

inside red cells is correlated with increases in the oxygen unloading potential (the Po_2 at a given % sat. and pH) attests to the widespread interest in oxygen

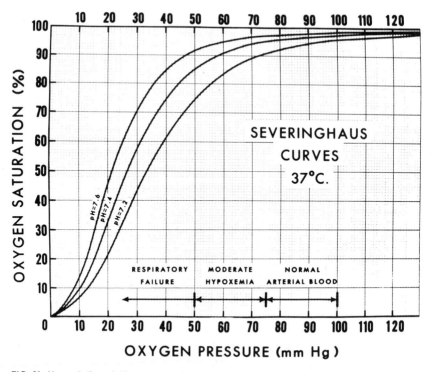

FIG. 31. Normal dissociation curves of man.[4]

transport among diverse investigators.[161–164] It has been found that increases in 2,3 DPG levels are associated with rightward shifts of the hemoglobin dissociation curve in man. Stresses of various kinds in normal men (e.g. ascent to high altitude, and exercise) and in patients (anemia, cardiac failure, myocardial infraction) produce transient (? permanent) rightward shifts and/or increased 2,3 DPG levels. Furthermore, patients with congenital heart disease and pulmonary patients with increased hematocrits have dissociation curves to the right of normal,[165] suggesting that an increase in Po_2 at a given % sat. and pH warrants being considered a "chemical reserve" among the mechanisms of oxygenation utilized by mammals.

The conventional measure of the oxygen unloading potential is the P_{50} (sometimes written T_{50} or $P_{1/2}$). This designates the Po_2 when the blood is 50% saturated with oxygen.

It is too early to be certain of all the physiologic advantages (and perhaps disadvantages) to man of rightward shifts in the dissociation curve. The circumstances under which transient shifts occur suggest that they are rapid adaptations to naturally occurring conditions of oxygen want and help to prevent tissue hypoxia.* A dramatic example of one advantage of a high O$_2$ unloading potential or high P$_{50}$ is seen in a report[155] of patients with "Hemoglobin Seattle." In two anemic patients the P$_{50}$ was 43.5 (normal 28.1); in the words of the authors, "low hemoglobin concentrations in the blood do not necessarily imply the presence of anemia in the physiologic sense . . . at least in states not associated with stress" because the rightward shift of the dissociation curve maintains a normal tissue Po$_2$. Because of the rapidity with which human data on blood O$_2$ intensity-quantity relationships are accumulating, we will examine better-established relationships in animals illustrating how the P$_{50}$ of blood can be considered in relation to other reserve mechanisms.

The sheep, which lives comfortably at 10,000 to 12,000 feet, has a hemoglobin concentration effectively about half that of man. Despite this, sheep supply their tissues with O$_2$ at the same rate and over the same Po$_2$ range as man. This is made possible because sheep blood (containing sheep hemoglobin B[167]) has a much higher O$_2$ unloading potential (P$_{50}$ = 41 mm Hg) than human blood (P$_{50}$ = 27 mm Hg). In Figure 32 it can be seen that sheep have a relatively right-shifted O$_2$ dissociation curve.

Lest, however, it be thought that a right-shifted curve is necessary to transport O$_2$ at high altitude, it should be remembered that the llama, who lives comfortably at 14,000 feet or more, has a left-shifted curve and a P$_{50}$ of only 21 mm Hg. At moderate altitude (1600 meters or about one mile above sea level, barometric pressure, P$_B$ = 635 mm Hg) the arterial Po$_2$ and mixed venous saturation, S$\bar{v}$O$_2$, are about the same for the llama and the sheep (about 75 mm Hg and 58% respectively, as seen in the upper left panel of Fig. 33). However, the llama's P$\bar{v}$O$_2$ (because of his left-shifted curve) is only about 26 mm Hg and the sheep would appear to have the higher tissue Po$_2$. Nevertheless, as the barometric pressure is progressively lowered, the sheep's arterial point sinks so low (beginning at 535, where it crosses the knee of the curve, to 335 where it is at the bottom of the steep portion), that its mixed venous point is forced to be on the lower flat portion (where P$\bar{v}$O$_2$ rapidly falls with lowered saturation toward 20 mm Hg and below). Thus the sheep cannot maintain O$_2$ transport at very high altitude while the llama can— largely because the llama, because of its left-shifted curve plus some hyperventilation can keep the arterial saturation from falling rapidly. With only a

* It is clear that in special "unnatural" conditions, a left-shifted curve can be an advantage. A classical example is the beneficial effect of CO in very severe hypoxia. As Haldane and Lorrain Smith showed,[166] a certain amount of added CO prevents animals from dying in very low O$_2$ atmospheres by helping blood to absorb O$_2$ in the lungs because of an increased affinity of hemoglobin as reflected by a left-shifted curve. Considerations of unnatural conditions are important in these days when artificial gas mixtures at high pressures, complete control of paralyzed patients by mechanical ventilators and hypothermic states can be imposed in a modern hospital.

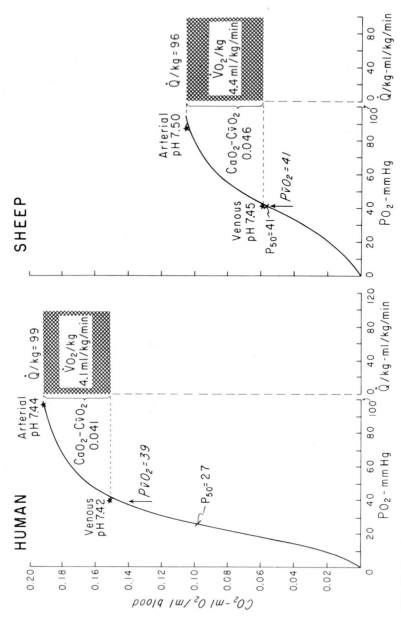

FIG. 32. The O_2 uptake per kilogram of sheep is comparable to that of man because although the O_2 quantity per liter of sheep blood is much lower, the O_2 intensity or unloading potential (P_{50}) is much higher than in man. (From Metcalfe,[142] 1970 with permission.)

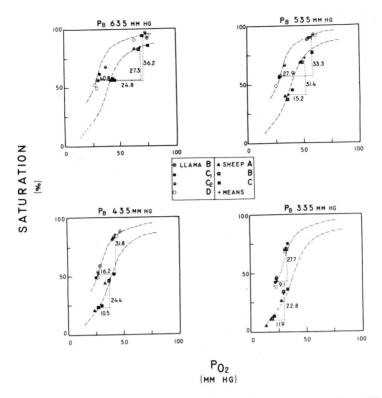

FIG. 33. *In vivo* hemoglobin oxygen dissociation curves in llama and sheep at four different barometric pressures. Individual values for % saturation and P$_{o_2}$ in arterial and mixed venous blood at each level of barometric pressure (P$_B$), as indicated above each panel, are represented by symbols (inset). Mean values for each species, at each altitude, are indicated by crosses and these values are connected by dotted lines. Numerals close to vertical dotted lines indicate arterial-venous oxygen saturation differences; numerals close to horizontal dotted lines indicate arterial-venous oxygen tension differences. The dashed lines are the best fitting oxygen dissociation curves for each species. The llama's curves are always to the left of the sheep's. (From Banchero *et al.*[168] with permission.)

slight increase in cardiac output, the llama can maintain adequate O$_2$ delivery[168] and can therefore keep the venous point on the steep portion of the curve as shown in the lower right panel of Figure 33.

This example illustrates how the adaptation of the llama to high altitude involves all of the first three reserves—circulatory, erythropoietic and chemical—and, probably the fourth or pulmonary reserve since its arterial P$_{o_2}$ remains higher than that of the sheep. To this pulmonary reserve mechanism we now turn.

4. The Pulmonary Reserve

The role of the lung as an oxygenator is limited to its capacity to maintain and increase, when necessary, the Po_2 (not the quantity) of oxygen in arterial blood. In health, at rest and with normal hemoglobin, the quantity of oxygen delivered to the tissues per unit time is more than adequate for tissue needs, but this delivery rate depends largely on the non-pulmonary reserves we have just reviewed. Two important clinical facts emerge from these considerations: (1) strictly speaking potentiometric arterial hypoxemia (low arterial Po_2) is the only disturbance of oxygenation attributable to lung failure and (2) the only strictly pulmonary reserve mechanism to combat hypoxia is to increase the pressure of oxygen in the blood. These facts can be illustrated by examining the differences between airway, shunt and diffusion abnormalities and by studying the way the pulmonary ventilation is adjusted in response to high altitude and in various diseases.

If the lung were a perfect oxygenator and had a single alveolar oxygen pressure, all blood leaving it would attain this pressure. The normal human lung at sea level fails to realize this degree of perfection (shown in Fig. 34) because arterial blood actually has a Po_2 about 10 mm Hg lower than alveolar. Arterial hypoxemia can occur from overall hypoventilation in which every alveolus is underventilated; the fall in arterial Po_2 is seldom as great as the 20 mm Hg shown in Figure 34 because the (necessarily) accompanying Pco_2

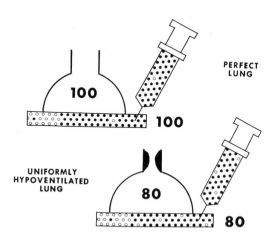

FIG. 34. The Po_2 in a normal alveolus at sea level is 100 mm Hg. Were the lung a perfect oxygenator, the blood leaving such an alveolus would have a Po_2 of 100 also. With normal hemoglobin affinity, this Po_2 converts nearly all Hb (open circles) to HbO_2 (closed circles).

The Po_2 in a hypoventilated alveolus is lower than 100 and the blood leaving it is consequently less saturated.

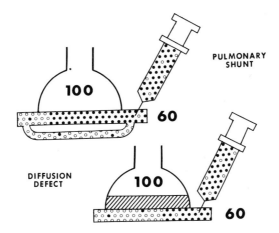

FIG. 35. A pure pulmonary shunt lowers arterial Po_2 but not in strict proportion to the shunt because shunting *per se* lowers only O_2 concentration. The Po_2 lowering depends on the steepness of the dissociation curve between the arterial and venous points; the steeper the curve the less the Po_2 drop.

A pure diffusion defect affects O_2 transport more than CO_2 transport because CO_2 diffuses through thickened tissues about 20 times more readily than O_2.

rise ordinarily drives the respiration and prevents severe hypoventilation from occurring.

Far more insidious, because arterial Pco_2 is so slightly raised by them, are abnormal anatomic shunts and diffusion impediments—insidious because, in general, arterial Po_2 must fall much more than the Pco_2 rises to evoke a hyperventilatory and hence protective response. These disturbances are depicted in pure form in Figure 35; they can actually occur in more or less pure form in, for example, diseases such as congenital pulmonary arteriovenous fistula and interstitial pulmonary granulomas and fibroses. The main reason why the arterial Pco_2 is low or normal in these conditions is primarily physicochemical as will become clear in the next chapter. Here we will focus on the effectiveness of the major pulmonary reserve mechanism (hyperventilation) in overcoming the three classical pulmonary lesions leading to arterial hypoxemia: pure airway obstruction, shunting, and diffusion block.

Consider pure airway obstruction. Obviously, a low alveolar and arterial Po_2 from this cause might be overcome by hyperventilation, that is, if the consequences of such hyperventilation were not too costly. However, as indicated in Chapter 1, when the metabolic cost of breathing is high enough, neither normal subjects nor patients "choose" to hyperventilate. In chronic hypoxic patients the "choice" is in part indicated by attenuation of the hypoxic ventilatory control (commonly seen in chronic airway obstruction) and has been appropriately termed "a compromise adaptation."[169] (Under what circumstances of clinical hypoventilation and, particularly, *how* should assisted breathing therefore be imposed? In the case of patients who "fight the machine" the observation[170] that the work of breathing may be higher with IPPB than without is clearly pertinent. See Chapter 6, Case 5.)

Consider a pure shunt. Here general hyperventilation, since it only adds oxygen directly to pulmonary *capillary* blood, is of almost no use in overcoming hypoxemia. This follows because capillary blood is normally "on the flat part of the dissociation curve" (nearly saturated with oxygen) so that raising its Po_2, which is all the lungs can do, increases this saturation very little. Patients grossly cyanotic because of right-to-left shunts of congenital heart disease often do not hyperventilate and, if they do, it may be, as Otis has observed,[171] more of an adjustment maintaining a near normal acid-base state than one which improves O_2 delivery to the tissues. In most cases of respiratory failure, such pure shunting is not present but rather a so-called "physiologic shunt" caused by atelectasis, interstitial edema, etc. Many of these patients can be helped by artificial ventilation and by giving oxygen. (Under what conditions are these measures indicated? How high a concentration of oxygen is needed? See Chapter 6, Case 6.)

Consider diffusion impediments. Both in relatively pure cases (pulmonary fibrosis) and in acute cases complicated by what may be called physioanatomic lesions (atelectasis and pulmonary edema, interstitial or alveolar), patients hyperventilate. This hyperventilation is of some advantage in these patients if it raises alveolar Po_2 since this raises pulmonary capillary Po_2. But overbreathing has many causes and many consequences (see Chapter 6). In the chronic cases in patients with diffusion impediments, for example, hyperventilation is associated with both persistent arterial hypoxemia and disabling dyspnea; the hyperventilation is not abolished when hypoxemia is relieved by oxygen breathing. This persistence of hyperventilation while breathing oxygen also occurs in association with many acute pulmonary disturbances and produces potentially dangerous alkalemia.[172] (See Chapter 6, Case 8.)

In most clinical conditions causing arterial hypoxemia the above three lesions are combined rather than separate. Only by special methods can their individual contributions to hypoxemia be assessed in living patients. Even though their effects are difficult to separate, it should not be concluded that they do not exist.

The fact that these three disturbances are often combined and associated with complicated airway abnormalities which disturb the ventilatory process has led to two extreme views about them and about hypoxemia of pulmonary origin in general. One view holds that diffusion difficulties do not exist clinically and that virtually all arterial hypoxemia is the result of regional ventilation-to-perfusion (V/Q) inequalities; the other holds that pulmonary shunting (by implication, *pure* shunting) adequately denotes any pulmonary contribution to hypoxemia without CO_2 retention. Briscoe, with his colleagues,[173] has shown that both a middle ground and precision are still possible in this field by quantifying the separate contributions of diffusion abnormalities, shunts and V/Q disturbances in patients with alveolar capillary block. In the light of increasing knowledge of the interstitial water space of the lung, their analysis of the diffusion impediment resulting from interstitial edema has obvious therapeutic implications. (See Chapter 6, Case 6.)

Nevertheless, it has been sometimes argued that even critically ill patients with profound hypoxemia have normal total "alveolar ventilation" if their Pco$_2$ is normal. It is also sometimes claimed that in such patients no pulmonary capillary blood is separated from alveolar gas by a membrane or tissue space thick enough to block O_2 transport (causing complete alveolocapillary block and hence a shunt) or slow down O_2 diffusion (and cause a partial alveolocapillary block) until proved otherwise. This argument is an extrapolation from an impressive and elaborate quantitative formulation,[174,175] namely, the theory of ventilation-perfusion ratio disturbances which explicitly excludes pure shunts and diffusion impediments. The extrapolation is to equate "physiologic shunting" with any hypoxemic disturbance defined by the V/Q theory.

This theory is illustrated by the lung model shown in Figure 36. The essential part of the theory for the clinician is that if a portion of the lung is not ventilated in proportion to its blood flow, the Po$_2$ in the blood leaving that portion will be relatively low compared with blood leaving a well-ventilated portion, but the saturation will be extremely low. Local hypoventilation, in brief, produces a drop in O_2 intensity in proportion to the severity of the hypoventilation, but a drop of O_2 quantity *out of* proportion to this severity. The mixed arterial blood O_2 content will thus be lowered and hence its O_2 tension as well. The lowering of content will be the more severe the greater the difference between the slopes of the upper (arterial and pulmonary capillary) portion of the O_2 dissociation curve and the middle and lower (venous) portion.

The well-known response of normal pulmonary vessels to alveolar gas hypoxia (probably vasoconstrictive)[176,177] may, of course, modify such distribution abnormalities in disease. In fact, if the pulmonary blood flow to a suddenly hypoventilated alveolus could be lowered sufficiently, the hypoxemic effect of a V/Q disturbance could be completely obviated by compensatory local vasoconstriction. When considering patients with chronic bronchitis and uneven ventilation whose hypoxemia is profound, it can be argued that such secondary pulmonary vascular compensation is absent and that the patient is a pulmonary vascular non-reactor.[178] In view of the tremendous variability of the pulmonary hypertension observed in normal persons at high altitude,[179] such an explanation for the chronic hypoxemia in these patients is reasonable.

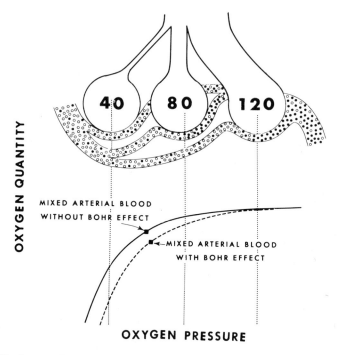

OXYGEN QUANTITY

MIXED ARTERIAL BLOOD
WITHOUT BOHR EFFECT

MIXED ARTERIAL BLOOD
WITH BOHR EFFECT

OXYGEN PRESSURE

FIG. 36. The hypoventilated alveolus with a Po_2 of 40 represents one third of a pathological lung with mismatching of ventilation and perfusion. The mixed venous blood flowing by this alveolus picks up little if any O_2 and the blood leaving it stays venous in character and has a very low saturation. The middle alveolus contributes a normal quantity of O_2 to its blood. The hyperventilated alveolus with a Po_2 of 120, however, contributes little more than the normal quantity of O_2 despite its high O_2 intensity. A right-shifted curve might be a disadvantage to patients with unevenly ventilated lungs because it interferes with the quantity of O_2 loaded even though it (slightly) increases the Po_2 of the mixed arterial blood.

Unequivocal evidence for pure regional V/Q changes (without shunts or diffusion defects) is extraordinarily difficult to obtain even with modern instruments applied to conscious cooperative patients. In patients acutely hypoxemic following non-thoracic trauma, deep anesthesia, myocardial infarction or pulmonary embolism (all conditions in which the parenchyma and vessels of the lungs are involved as much as or more than the airways), the V/Q theory uncontaminated by unmeasurable effects caused by pure shunting from atelectasis and pure diffusion block from edema is even more difficult to support with evidence.

Is it fortunate that as the V/Q theory was developing, comparable sophistication in the field of pulmonary mechanics was also developing[180-182] because successful treatment of arterial hypoxemia, especially that accomplished with the aid of mechanical ventilators, depends as much on an understanding of the pulmonary tissue changes of patients in respiratory failure as on understanding the physical chemistry of the process of mixing blood samples at different gas tensions (which is the basis of the V/Q theory). In fact, with increasing sophistication in mechanical respirator therapy, it is becoming important to consider how much hypoxemia is caused by the three classical disturbances (overall hypoventilation in relation to metabolism, pure shunting, and alveolar interstitial edema interfering with oxygen diffusion). The importance lies in the fact that different patterns of imposed breathing seem to be suitable for different causes of hypoxemia (see Case 6, Chapter 6).

How is the pulmonary reserve used in oxygenation disturbances? To what demands does the pulmonary ventilation respond? If the demand is considered to be the maintenance of a normal arterial Pco_2, the answer is fairly easy: ventilation is regulated by the feedback mechanism described in Chapter 1. If the demand is considered to be the maintenance of a normal arterial pH, the answer is more difficult as shown by considering how "compensations" for metabolic acidosis and alkalosis are achieved (Chapters 3 and 5). If the demand is muscular exercise, the answer in detail is not available despite many years of study; however, what can be said is that the ventilation is (for moderate exercise) a *linear function of the O_2 uptake*,[183] i.e., the ventilatory supply mechanism is somehow adjusted to O_2 demands.

Since the ventilation directly affects arterial Po_2, it might be supposed that one demand might be "to maintain a normal arterial Po_2." Under special circumstances (e.g. with alveolar and arterial Pco_2 held constant), ventilation is indeed a function of arterial Po_2, increasing (non-linearly) with progressively severe potentiometric hypoxemia* induced in normal subjects breathing low oxygen mixtures while comfortably seated at rest. However, circumstances in which blood gases are not controlled and the subject (or patient) is not at rest are very different. For example, when Po_2 and Pco_2 are allowed to vary naturally during exercise, (along with all other factors controlling ventilation) the effect of the intensity of arterial oxygenation (PaO_2) seems to be completely submerged in comparison to the overwhelming demands for oxygen *in quantity per unit time*.

A dramatic example of the overriding influence on ventilation of demands for oxygen quantity as contrasted to the demands for maintaining arterial blood gas levels was provided by Reeves *et al.*[186] who collected and/or analyzed measurements made in men exercising at four different altitudes. Regardless of the altitude (and hence the inspired Po_2), the ventilatory response (see V_E in Fig. 37) was the same in relation to relative oxygen demands, that is, to how

---

* This "ventilatory response to hypoxia" can be considered to be perhaps determined by tissue oxygen tension as affected by oxygen delivery rather than by arterial Po_2 since the response is linearly related to arterial O_2 saturation or content.[184]

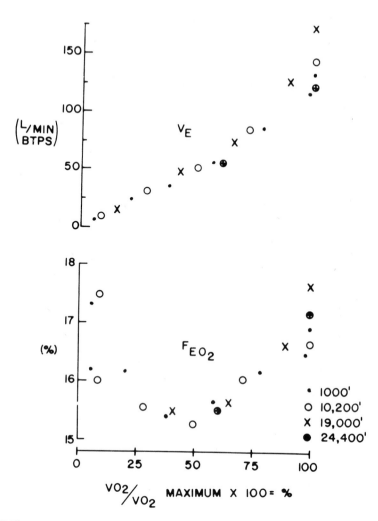

FIG. 37. The upper panel shows how healthy young men (ages 15 to 35 years) use their pulmo-nary reserve (increase of ventilation) in response to exercise at different altitudes as plotted against the ratio: actual O_2 uptake/maximal O_2 uptake. The abscissa (relative O_2 uptake) measures how much of the total capacity to transport O_2 is being utilized. (See Astrand and Rodahl.[185])

   The lower panel indicates that a minimum of wasted ventilation (in terms of gas phase O_2 transport) is reached at moderate exercise because F_{EO_2} (and hence the O_2 ventilation equivalent or O_2 v) is minimal at that point. (Adapted from Reeves et al.[186])

much O_2 was needed to perform the exercise relative to the total capacity to transport oxygen. "Presumably the working muscles themselves best sense how closely they approach maximum effort and transmit this information to the central nervous system. The accumulation of some metabolite such as lactic acid, either in the working muscles themselves or in blood borne to the chemoreceptors, could provide a stimulus for increasing ventilation. The present investigation emphasizes the importance of examining regulatory mechanisms which provide a unified concept of ventilation control during exercise at all altitudes available to men."[186]

Such circumstances are in some ways more like those of an intensive care unit (where "all other factors" cannot be controlled) than those of a laboratory investigation of the control of breathing in an anesthetized quadruped on his back. Thus both in physiologic and clinical work it seems worthwhile, to "make some estimate of the call [for oxygen] which the blood has to meet, before entering into a discussion of how the call is to be met."[103]

MEASUREMENT OF OXYGENATION INDICES AT THE BEDSIDE

Arterial versus Venous or Capillary Blood

While capillary and even venous blood can give useful information with regard to Pco_2 and pH, it is becoming clear that when the state of oxygenation of a critically ill patient is at issue, an arterial sample is required to obtain a generally useful Po_2 measurement.[187,188]

Both the Po_2 and a colorimetric determination of $\% O_2$ saturation can be determined in a few minutes with modern instruments. The accuracy of both determinations depends on how well the instruments are cared for and how carefully they have been calibrated. Two monographs on oxygen measurements in biologic systems have recently been published.[189,190]

Po₂

The development[191] of a platinum electrode separated from an external medium by a membrane permeable to oxygen but impermeable to water and ions constitutes the major technical advance in biologically applicable oxygen measurements. The availability of oxygen molecules at the platinum surface is proportional to the passage of current resulting from a given potential applied to the platinum electrode.

The Clark electrode is satisfactorily and easily calibrated, using analyzed humidified gas at the temperature of the electrode.

Although devised originally to measure the Po_2 in fluids, the Clark electrode can be used to measure the Po_2 of inspired and expired gas. Because it is very temperature dependent, the accuracy of continuous measurement of respired gas depends on the care with which temperatures are controlled.[190]

% Saturation

Spectrophotometric (colorimetric). Several "oximeters" are available for determining the ratio of the oxygenated to the total hemoglobin in a blood sample. This ratio is determined from the ratio, at two specific wavelengths, of the quantity of light which is transmitted through or reflected from a blood sample. Modern reflectance instruments are adequate means of measuring oxygen saturation for clinical work. No oximeter can give a direct indication of the quantity of oxygen in blood.

Oximeters are best calibrated by comparing their results with those of Van Slyke analysis. Under most conditions, colorimetric results are accurate to within 2 to 3% saturation units. (Less accuracy is to be expected for venous than for arterial blood.)

Van Slyke analysis. The oxygen in an anaerobically collected blood sample is extracted and measured[56] as well as the O_2 capacity of an aliquot sample equilibrated with room air (or oxygen sufficiently high to saturate hemoglobin). With corrections for physically dissolved oxygen,[108] the ratio of the bound oxygen (O_2Hb) in the sample to that in the aliquot (sum of O_2Hb and Hb) determines the % saturation. This is the most reliable method yet devised. Its only objection is the time and skill required for analysis.

"Calculated saturation" from a dissociation curve. A point located on a standard dissociation curve[4] by accurately measuring both Po_2 and pH would give an accurate "calculated" % saturation. The exact position of the curves of patients and even of normal persons makes this method somewhat inaccurate (especially on the steep portion of the curve).

O_2 Content

The O_2 content of a sample of blood can be calculated (if the hemoglobin is assumed to have a normal O_2 affinity) from the formula

$$O_2 \text{ content (cc/L)} = 1.34 \times [Hgb] \text{ (in g/liter)} \times \% \text{ saturation}/100.$$

THE CLINICAL SIGNIFICANCE OF THE ARTERIAL Po_2

Because of the fact that at or near sea level arterial blood points on the oxygen dissociation curve lie on the upper "flat" or nearly horizontal part, and because the lungs only directly affect the Po_2 of pulmonary capillary blood (and none of the other indices of oxygenation), the arterial Po_2 is a sensitive indicator of defective function of the lung parenchyma. A low arterial Po_2 may be one of the earliest clinical measurements indicating the presence of pulmonary embolism, atelectasis and pulmonary infiltrations of many kinds. Potentiometric hypoxemia is, therefore, frequently a sign which portends progressive respiratory failure.

A low PaO_2 is not, however, a sign that the oxygen content of the blood is low or that oxygen delivery to any tissue is deficient. In itself, then, potentiometric hypoxemia need not produce any physiologic disturbance, depending largely on the extent to which circulatory, erythropoietic and chemical reserves have compensated for the pulmonary disturbance usually present.

In summary, the PaO_2, considered apart from other blood gas measurements (especially the Pco_2), and from clinical signs reflecting the oxygenation of tissues, has only *general* clinical significance. To be specifically useful in diagnosis and treatment it must be related to other blood gas measurements, to reserve mechanisms which influence Po_2 in different parts of the body (see Table 9) and, most important of all, to the total clinical picture.

TABLE 9. *Site of major effect in respect to type of O₂ transport reserve*

Type of Reserve	SITE OF MAJOR EFFECT			
	Arterial	Capillary	Venous	Tissue
Pulmonary..........	PaO_2			
Chemical............	(% Sat)..........	$P\bar{c}O_2$		
Erythropoietic.......	(CaO_2)...........................		PvO_2	
Circulatory..........	(Delivery)..			P_TO_2

The lungs directly affect only arterial Po_2. The % saturation and the position of the dissociation curve are determined by the chemical affinity of hemoglobin for oxygen. The chemical reserve primarily affects the mean Po_2 in the peripheral capillaries. Increased erythropoiesis, for a given % saturation, cardiac output and O_2 demand, raises CaO_2 and hence CvO_2 and PvO_2. Finally, although all reserves contribute indirectly to the maintenance of P_TO_2, the adequate delivery of O_2 ($CaO_2 \times \dot{Q}$) is the most powerful reserve for preventing tissue hypoxia.

The four arterial indices of oxygenation are listed in the usual order from top to bottom. This order allows the "cascade" of Po_2 levels in the body to be displayed. Compare Table 5.

5

···

Blood Gas Interactions and
Acid-Base Disturbances

"Every adaptation is an integration."

JOSEPH BARCROFT (1938)[159]

···

IT HAS BEEN our aim so far to study P_{CO_2}, $[HCO_3^-]$, pH and P_{O_2} separately since each has inherent significance. As chemical measurements they can be understood in isolation; as physiologic variables they cannot be; in medicine they must be connected in the physician's mind since they are integrated in the patient's body.

Arterial blood gas data are conveniently grouped in the following way:

pH . . . . $[HCO_3^-]$ . . . .
P_{CO_2} . . . .   BE   . . . .
P_{O_2} . . . .   % Sat   . . . .

The meaning of BE (Base Excess) is discussed in Chapter 2 and in Appendix III and the % saturation defined and considered in Chapter 4. In this and the following chapter we will summarize how the first four variables listed are linked in respiratory, renal, circulatory and cerebral processes and show how blood gas measurements can be used in diagnosis and treatment.

The four classical pure primary acid-base disturbances are listed in Table 10. As actually encountered clinically, acid-base abnormalities are often mixed (i.e. resulting from more than one primary disturbance). Therefore, in addition to defining each of the four primary disturbances in chemical terms and indicating how secondary compensation is achieved, we will illustrate by

111

TABLE 10. *The classical pure primary acid-base disturbances and their compensatory process*

Primary disturbance	Compensatory process
PRIMARY HCO$_3^-$ DEPLETION (Metabolic acidosis)	SECONDARY CO$_2$ DEPLETION (~~Respiratory alkalosis?~~)
PRIMARY HCO$_3^-$ RETENTION (Metabolic alkalosis)	SECONDARY CO$_2$ RETENTION (~~Respiratory acidosis?~~)
PRIMARY CO$_2$ DEPLETION (Respiratory alkalosis)	SECONDARY HCO$_3^-$ DEPLETION (~~Metabolic acidosis?~~)
PRIMARY CO$_2$ RETENTION (Respiratory acidosis)	SECONDARY HCO$_3^-$ RETENTION (~~Metabolic alkalosis?~~)

The least ambiguous terms are in capital letters. A compensatory process tends to return the pH toward normal. In parentheses are the usually employed terms. It is convenient to make these terms stand only for primary processes. Hence they are crossed out on the right to indicate the inconvenience of their standing for compensatory processes.

clinical examples how two or more primary disturbances can co-exist and either exaggerate or minimize pH changes. Such complicated cases are worth studying because they "demonstrate the need to use both clinical and laboratory observations in making an acid-base diagnosis" and "to know the expected limits of a secondary compensatory response" because "only in this way can simple primary and mixed primary disturbances be separated."[97]

Although no attempt is made to deal completely with fluid and electrolyte imbalances, certain important connections between potassium levels and blood gas disturbances will be discussed.

PRIMARY [HCO$_3^-$] DEPLETION WITH AND WITHOUT ACIDEMIA

Clinical Description

Patients whose bicarbonate reserve has been *acutely* lowered by fixed acids vary from those showing only the mild hyperpnea of early diabetic ketosis to those in coma from the lactic acidosis and hypoxia caused by traumatic shock. Symptoms and signs of metabolic acidosis arise mainly because of lowered pH in the blood, the myocardium and the brain: hyperventilation, cardiac dysrhythmias and many forms of cerebral and brain-stem dysfunction are common.

The irritability and apathy of the dehydrated infant with diarrhea, the dull delirium associated with uremic encephalopathy, the tachycardia and circulatory failure of the severely injured are among the manifestations of primary [HCO$_3^-$] depletion. Hyperventilation (Kussmaul breathing) is very common; however, with thoracic trauma or CNS depression (as in methyl alcohol poisoning) this helpful diagnostic sign may be absent.

Chronic acidemia (as in renal disease, when the blood pH stabilizes at a constant "subnormal" level, usually 7.25 to 7.35) rarely causes symptoms except for those arising from decalcification of the bones.

Chemical Description

Fixed (non-carbonic) acid in excess, being mainly buffered by bicarbonate, lowers [HCO$_3^-$] and produces metabolic acidosis; loss of base (as in diarrhea or by intestinal or biliary fistulae) also depletes the body of bicarbonate.

The major reactions in the plasma are outlined as follows:

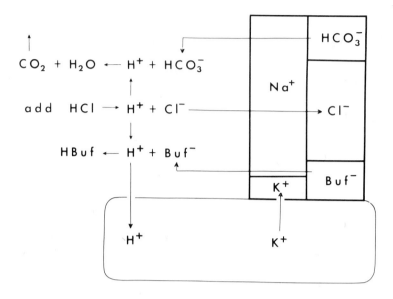

The invading fixed acid is HCl in this formulation. Its protons are mainly bound by plasma HCO_3^-, because the lungs provide a vehicle for the escape of CO_2. Blood buffers (Buf^-) also contribute to the temporary* sequestration of H^+ as do intracellular H^+ binders. Potassium tends to leave the cell and enter the plasma. In diabetic ketosis a large amount of K^+ is lost by osmotic diuresis; the resulting dehydration raises serum $[K^+]$ so that only after therapy (insulin, fluid replacement, etc.) does the potassium depletion become evident by a fall in serum $[K^+]$.[192]

In summary, an invasion of an acid like β-hydroxybutyric or lactic acid immediately reacts with plasma bicarbonate. Although CO_2 is driven from the body, the Pco_2 fall is seldom in proportion to the bicarbonate decrease. Hence pH falls:

$$\downarrow pH = 6.1 + \log \frac{[HCO_3^-]\downarrow\downarrow}{.03\ Pco_2\downarrow}$$

Means of Compensation

The low pH of the blood is sensed both by the peripheral and central chemo-receptors and hyperventilation results (e.g. as in the Kussmaul breathing of diabetes). In addition, normal kidneys, by increasing urinary titratable

---

* Eventually the excess H^+ of fixed acids like HCl must be excreted by the kidneys.

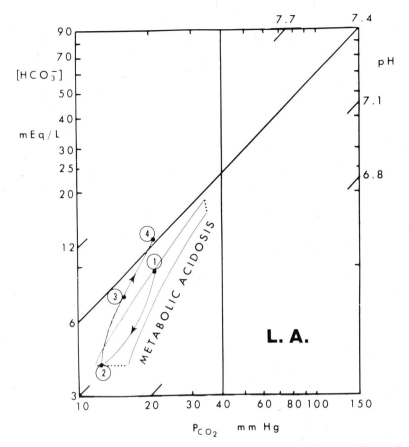

FIG. 38. Two of the four points of Case 1 fall within the expected region or significance band for metabolic acidosis—at least as established for children.[193] For any measurement within this region or band, the expected amount of compensatory hyperventilation has lowered the P_{CO_2} below 40 mm Hg enough to return pH *toward* 7.4.

acidity and NH_4^+ excretion, can remove as much as 500 mEq of unwanted acid per day. How much respiratory compensation is to be "expected" is indicated by the "significance band" shown in Figure 38. Whether or not such bands have practical significance will be discussed later.

Clinical Causes

Bicarbonate depletion is historically the first of the four primary classical acid-base disturbances to have been recognized and is the best understood. For this reason its commonest causes will simply be listed: diabetic ketosis, cellular hypoxia (as in shock which may lead to lactic acid accumulation), diarrhea, renal failure and severe muscular exercise. It should be remembered that plasma $[HCO_3^-]$ may be low for reasons other than that of fixed acid invasion. (See Chapter 2.)

> *Case 1.*   Metabolic acidosis before and after compensation.
> L.A., a 32 year old diabetic with Kimmelstiel-Wilson nephritis, was admitted to the hospital because of a "cold." Twenty-four hours before admission he had vomited and noticed difficulty in breathing. P.E.: T, 37°C; P, 100; BP, 150/115; R, 24. Breathing deeply, tongue parched, pharynx dry. Fundal hemorrhages and microaneurisms. Initial electrolytes: Na 124, Cl 88, K 3.4, CO$_2$ content 9. During the first 24 hours, only fluids and one ampule (44 mEq) of Na HCO$_3$ were administered. The pertinent arterial blood gases were:

> ①   pH   7.285 $[HCO_3^-]$ 9.5
> P_{CO_2} 21   BE   −16

> The patient was treated with insulin and saline solution, but felt "uncomfortable" all that day. On the third hospital day he was found to have frequent premature ventricular contractions, his serum K was 3.1 and the blood gases were:

> ②   pH   7.15 $[HCO_3^-]$ 4.0
> P_{CO_2} 12.5   BE   −25

> Within a few hours of more aggressive HCO$_3^-$ and potassium replacement, the dysrhythmia disappeared and the patient felt less dyspneic. The afternoon and evening blood gases were:

> ③   pH   7.31 $[HCO_3^-]$ 7.5
> P_{CO_2} 15.5   BE   −17.5
> ④   pH   7.40 $[HCO_3^-]$ 12.5
> P_{CO_2} 20.5

> Comment: This was a case of pure (unmixed) primary HCO$_3^-$ depletion, i.e. "metabolic acidosis." Despite considerable hyperventilation, on admission the patient had not returned his plasma pH to normal. It was 7.285. Although this was an unmixed acid-base disturbance, there was some (incomplete) P_{CO_2} lowering by secondary hyperventilation. (This is sometimes called secondary "respiratory alkalosis"—a confusing term when pH is low; see Table 10, top line.)
> Delay in treatment allowed the pH to fall to 7.15. Points ① and ② fall within the expected band of chronic metabolic acidosis (Fig. 38). Bicarbonate therapy raised $[HCO_3^-]$ sufficiently to bring the pH to 7.31 and finally to 7.40. Therefore points ③ and ④ fall above the significance band.

PRIMARY HCO$_3^-$ RETENTION WITH AND WITHOUT ALKALEMIA

Clinical Description

Bicarbonate retention, unless associated with severe alkalemia (pH > 7.55), with hypokalemia, or with hypoxemia, has few clinical manifestations.

"Delirium and obtundation due to metabolic alkalosis are rarely severe and never life threatening."[129] Sudden pH changes caused by an increase in HCO_3^- are far less likely to affect cellular function than when induced by changes in Pco_2 (see Fig. 48). The diagnosis depends largely on a history of alkali ingestion or evidence of acid loss via gastrointestinal or renal routes.

Low serum potassium, which often accompanies HCO_3^- retention, can cause many neuromuscular disturbances (involving smooth, skeletal and cardiac muscle), impaired gastric secretion, altered renal function, EKG changes and cardiac arrest.

If a secondary Pco_2 increase occurs (see below), alveolar Po_2 may fall enough to cause arterial hypoxemia. Severe disturbances referable to oxygen lack are unusual in metabolic alkalosis.

Chemical Description

Plasma $[HCO_3^-]$ is increased either by simple addition or by *generation* of bicarbonate, the latter occurring when fixed acid is lost from the body. The major reactions are shown as follows:

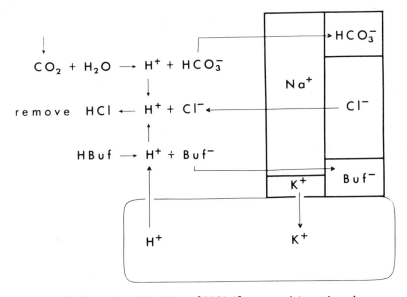

The alkalemia produced by loss of HCl (for example) tends to lower extracellular $[H^+]$. Protons tend to leave the cells and potassium to enter them.

This figure, like that on page 114, represents an attempt to show how a dynamic process occurring in time (on the left) leads to changes in the distribution of material in plasma (on the right). The units on the left are mEq per unit time; on the right the units are mEq per liter.

When [HCO_3^-] is high, extracellular [K^+] tends to be low (Darrow's phenomenon), but which change is primary is often not clear. Hypochloremia may perpetuate HCO_3^- retention by interfering with its excretion by the kidney.[194] Diuretics may aggravate the associated alkalemia by causing "the extracellular space to contract around the initial extracellular bicarbonate pool" and hence produce "contraction alkalosis."[71,97] That alkalemia may result from contracting the extracellular bicarbonate space and hence concentrating the HCO_3^- can be seen in the following equation:

$$pH\uparrow = 6.1 + \log \frac{\text{concentrated } HCO_3^- = \uparrow[HCO_3^-]}{0.03 \times \text{unchanged } P_{CO_2}}.$$

Means of Compensation

Although CO_2 retention by the lungs is never very marked, some compensatory hypoventilation may occur (which raises P_{CO_2} and tends to lower pH toward normal). Such hypoventilation does not occur when K^+ losses are so great that H^+ enters cells and lowers intracellular pH, which apparently prevents respiratory depression,[100] as previously discussed on page 69.

No "significance bands" have been published.

Clinical Causes

Many if not most cases of HCO_3^- retention are iatrogenic and result from the use of diuretics, low salt diets, gastric lavage and/or unappreciated potassium depletion. HCO_3^- retention is the most complicated of the classic acid-base disturbances because of the associated ionic shifts, especially of K^+.

Loss of acid from the body. (a) Vomiting. An obvious cause of acid loss is vomiting due to duodenal obstruction. (Vomiting in biliary tract disease may result in the loss of more base than acid.) Gastric suction can remove surprisingly large amounts of HCl and cause profound alkalemia in a day or two. Though K^+ is also lost in vomitus, more is often lost in the urine.

(b) Hypokalemia. Acid loss in the urine frequently accompanies hypokalemia—the so-called paradoxical aciduria of potassium depletion.

Increased base intake. (a) Base ingestion. Alkalemia frequently occurs with sodium bicarbonate ingestion. Non-absorbable antacids taken by mouth do bind H^+ to some extent and lead to alkalemia but the excess base is normally excreted by the kidney so that, although the urine becomes alkaline, the blood pH is scarcely affected.

(b) Intravenous fluids containing base or base-forming salts (e.g. Na lactate). The current vogue of administering sodium bicarbonate for bronchospasm, cardiac dysrhythmias and many circulatory abnormalities has made iatrogenic alkalosis a common occurrence.

Potassium depletion. The most common form of hypokalemia is physician-induced via the use of diuretics. Compared with the renal sodium-conserving mechanism, the kidney's ability to conserve potassium is poor. Other factors leading to potassium loss are: a reduced dietary potassium intake, osmotic diuresis (as in diabetes), certain injuries such as burns, potassium-losing nephritis and the effect of corticosteroids, especially the mineralocorticoids.

Case 2. Metabolic alkalosis added to respiratory acidosis; two primary disturbances occurring sequentially.

R.P., a 27 year old mild asthmatic, caught a cold, developed mild chest tightness and was given an unknown antibiotic. This increased her chest tightness and after three days she was admitted to the hospital having received no other drugs. T, 37.5; P, 110; BP, 110/70; R, 30. She was moderately dyspneic and very frightened. The lungs were clear except for wheezing rhonchi. Electrolytes were not measured but blood gases were:

①	pH	7.32	$[HCO_3^-]$	25
	P_{CO_2}	50	BE	+1
	P_{O_2}	94	% Sat	95

The pH, P_{CO_2} and $[HCO_3^-]$ are plotted at point ① in Fig. 39.

The patient did not improve on antibiotics and aminophylline and her apprehension increased. She was started on corticosteroids. On the third day she became much more dyspneic and began to fatigue. Acute CO_2 retention was revealed by the following blood gases:

②	pH	7.20	$[HCO_3^-]40(\Delta[HCO_3^-] = +16)$
	P_{CO_2}	105	BE + 6

An emergency tracheotomy was necessary after an unsuccessful attempt to ventilate the patient by intubation. During these procedures she was given 6 ampules of $NaHCO_3$ (264 mEq) intravenously because of sudden supraventricular tachycardia and the threat of cardiac arrest. On day 4, with the patient on a mechanical ventilator, pH and P_{CO_2} were recorded in the notes but HCO_3^- retention was not explicitly recognized.

③	pH	7.46	$[HCO_3^-]50(\Delta[HCO_3^-] = +26)$
	P_{CO_2}	70	BE + 20

Fifty cm of inspiratory positive pressure were required to ventilate the patient. "Fluid overload" was suspected and 2 cc of Mercuhydrin were given. Gastric distention developed and a nasogastric tube was inserted. The gastric aspirate contained Cl^- 126 and K^+ 12 mEq/L. Serum electrolytes day 5: Na^+ 138, K^+ 2.7, "CO_2" > 40. Patient weaker, still obstructed. She was given acetazolamide and furosemide.

At 8:00 A.M. on day 5 the blood gases were:

④	pH	7.63	$[HCO_3^-]59(\Delta[HCO_3^-] = +35)$
	pCO_2	56	BE + 30

and a 24 hour urine was found to contain 132 mEq of potassium.

When the severity of HCO_3^- retention and K^+ depletion was recognized on day 5, KCl was infused at the rate of 25 mEq/hr and in a few hours the patient was

5

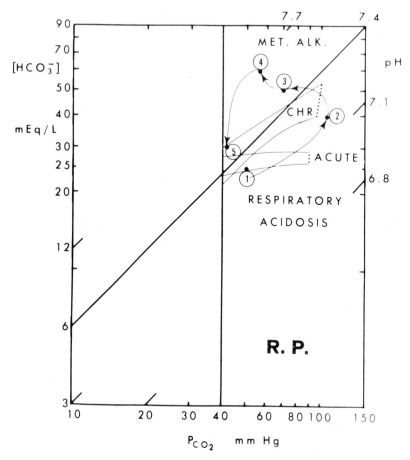

FIG. 39. Mixed respiratory acidosis and metabolic alkalosis. Point ① is in the acute hypercapnia band.[195] The chronic band is that of Brackett *et al.*[196] Point ② (sudden worsening) is beyond the range of both. Bicarbonate overdose and potassium depletion caused the metabolic alkaloses of points ③ and ④.

dramatically improved and made an uneventful recovery from day 6 on. On day 12 the blood gases were:

⑤   pH   7.48   [HCO₃⁻]   30
     Pco₂   41

The pertinent blood gas measurements are plotted in Figure 41.

Comment: The initial respiratory acidosis was mild. The patient's acid-base state, as shown by the position of point ①, was as expected in chronic asthma with a mild exacerbation. CO₂ retention became suddenly worse at point ② after which bicarbonate was given adding a primary metabolic alkalosis at ③. At point ④ the pH was 7.63 and [K⁺] 2.7 before KCl dramatically lowered the bicarbonate to point ⑤ the patient's usual state of (slightly over-compensated) chronic CO₂ retention.

It is interesting that the deviations of $[HCO_3^-]$ above normal were larger than the deviations of BE above normal. This is because BE is calculated from an *in vitro* CO_2 dissociation curve that does not take account of possible (probable in this case) migration of HCO_3^- from vascular to interstitial fluid in acute CO_2 retention. Thus the positive $\Delta[HCO_3^-]$ was, in this case, a better index of "base excess" than the BE.

PRIMARY CO₂ DEPLETION WITH AND WITHOUT ALKALEMIA

Clinical Description

Once thought to be nearly limited to neurotic patients with "psychogenic hyperventilation," primary respiratory alkalosis is now recognized as a common sign of serious disease. It should be thought of in caring for patients with unexplained agitation, and with bizarre symptoms such as circumoral paresthesias, with numbness and tingling of the fingers, in many acute poisonings (which often cause transient hyperventilation before depressing respiration), and in numerous cardiac and pulmonary disturbances (e.g. congestive heart failure, pulmonary atelectasis and embolism).

Chemical Description

In primary CO_2 depletion the arterial Pco_2 is low (below 35 at sea level) as a result of hyperventilation not secondary to an excess of fixed acid. Hence any $[HCO_3^-]$ reduction found in *unmixed uncompensated* (i.e. simple) primary CO_2 depletion is by definition secondary to a fall in Pco_2. The major reactions in the plasma are outlined as follows:

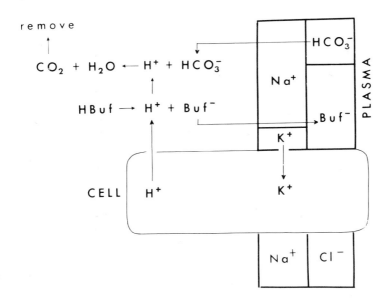

Hyperventilation for more than a few minutes leads to two reciprocal changes: plasma $[HCO_3^-]$ falls and $[Buf^-]$ rises, plasma $[K^+]$ falls and plasma pH rises. The plasma $[K^+]$ falls because it migrates from plasma to certain cells apparently in exchange for H^+.[119,198,199] The plasma pH rises because although H^+ enters plasma from cells, it does so at a rate lower than it leaves the plasma by reacting with HCO_3^-.

The primary change is the lowering of Pco_2 which pulls the top reaction to the left; HCO_3^- binds H^+ and some of it is decomposed. Hence $[HCO_3^-]$ becomes somewhat lowered but not enough to keep pace with the fall in Pco_2 (since H^+ ions are not supplied from extracellular buffers, H Buf, or from the cells in sufficient quantity). Therefore blood pH rises because of the slight degree of $[HCO_3^-]$ lowering compared with that of the Pco_2:

$$\uparrow pH = 6.1 + \log \frac{[HCO_3^-]\downarrow}{0.03\ Pco_2\downarrow\downarrow}$$

How much lowering of $[HCO_3^-]$ occurs in "simple" CO_2 depletion depends on how simple the system is from which CO_2 is being removed.* In 12 anesthetized patients artificially hyperventilated for two hours, the plasma $[HCO_3^-]$ decreased about 1 mEq/L for a 4-mm Hg decrease in Pco_2.[195] This is to be compared with the 1 to 5 ratio shown in Figure 7; part of the greater change in $[HCO_3^-]$ in CO_2 depletion than in CO_2 retention simply results from the fact that the CO_2 dissociation curve is steeper below $Pco_2 = 40$ than above.

Means of Compensation

When hyperventilation is maintained for more than a few hours, the plasma pH tends to return toward normal because the plasma $[HCO_3^-]$ becomes further lowered. Although the kidney is known to aid in this compensation by excreting bicarbonate,[50,200,201] this is not the only means of correcting the alkalemia. Lactic acid production, an almost invariable concomitant of hyperventilation, lowers plasma bicarbonate by reacting with it and producing CO_2. Thus, it has been written[202] that, "in addition to the renal excretion, bicarbonate loss can occur via the lungs."† Under some conditions, (e.g. hyperventilation during general anesthesia) more compensation may be achieved by this mechanism than by the kidney.

There are too few reports of lactic acid buffering in patients with primary

---

* Is it blood in a tonometer, or in the vessels of a normal volunteer, an anesthetized dog, or a patient with one or more primary disturbances? Blood in a tonometer has its $[HCO_3^-]$ lowered by CO_2 loss to a greater extent than blood in the body[45,197]; hence tonometer blood is better buffered and has a steeper CO_2 dissociation curve (see Chapter 2).

† This is metaphoric. The lungs excrete the CO_2 produced by the destruction of HCO_3^- by lactic acid.

CO_2 depletion to evaluate its clinical importance in defending *blood* pH in this disturbance; in the cerebrospinal fluid, however, lactic acid production appears to play a major role in defending CSF pH from the effects of hyperventilation.[203,204]

Clinical Causes

Psychogenic hyperventilation. Typically the patient is a young woman. Symptoms are dizziness, light-headedness and (rarely) tetany. Respiratory symptoms are "inability to take a deep breath," tightness in the chest, sighing, yawning—all often being called "shortness of breath" by the patient. The symptoms can be easily confused with those of serious organic illness, especially early diffuse pulmonary vascular disease.[205]

Iatrogenic hyperventilation. When this is deliberate (as when under the control of an anesthetist) and controlled (especially by monitoring end-tidal P_{CO_2} or arterial pH), serious acid-base disturbances can be avoided. When P_{CO_2} is rapidly lowered in the blood, however, it is rapidly lowered in the brain (because CO_2 is such a permeant molecule) and hence CSF pH can rise precipitously. Except when breathing high oxygen mixtures, cerebral hypoxia may result because the low P_{CO_2} not only produces cerebral vasoconstriction but, through the leftward shift of the hemoglobin dissociation curve, may interfere with the unloading of oxygen from the blood.

Patients whose P_{CO_2} is normally 60 to 80 mm Hg (e.g. in chronic obstructive pulmonary disease with CO_2 retention) are those in whom too rapid a lowering of this P_{CO_2} by mechanical ventilation is especially dangerous. The *arterial P_{CO_2} should not be lowered faster than 10 mm per hour* in these patients; otherwise convulsions (probably due to cerebral ischemia) and hypotension (cause not well understood) sometimes occur.[206,207] Acetazolamide, by promoting HCO_3^- excretion by the kidney (and perhaps by other actions in the central nervous system), may help to lower $[HCO_3^-]$ in proportion to the fall in P_{CO_2} and hence may lower pH in the blood (and in the environment of brain cells). (See comments on Case 5.)

Hypoxemia. As we have seen (p. 105), low oxygen levels in the arterial blood stimulate breathing. At altitude this hypoxic drive is partly offset by the fact that P_{CO_2} is lowered. In many patients with uncompliant lungs (especially those with interstitial disease and pulmonary embolism) this hypoxic drive is often accompanied by an additional reflex stimulation of respiration, probably via vagal afferent impulses from the lungs. In such patients, hyperventilation and a low arterial P_{CO_2} may persist for months to years.

Hepatic cirrhosis. Although primary CO_2 depletion with resulting high blood pH is fairly common in patients with hepatic cirrhosis, progressive liver failure leads to lactic acid accumulation which may drive the pH down to normal and below normal.[209] Hence, both alkalemia and acidemia can occur during the course of liver disease. The causes of both the hyperventilation and the increase in lactic acid are unknown. The mental symptoms of hepatic insufficiency have long been thought to be related to increased cerebral ammonia (P_{NH_3}) levels and these, in turn, to be further increased by increasing the pH by the equation

$$pH = 9.2 + \log \frac{K \cdot P_{NH_3}}{[NH_4^+]}.$$

However, exactly the opposite may be the case, since striking clinical improvement has recently been reported[210] to follow sodium bicarbonate administration, a finding of interest in view of the correlation between symptoms of pre-coma and the development of renal tubular acidosis.[211]

> *Case 3.* Respiratory alkalosis and metabolic acidosis occurring simultaneously.
>
> M.G., a 39 year old man with subacute lymphatic leukemia, developed high fever and pulmonary infiltrates. He underwent a lung biopsy which revealed disseminated coccidioidomycosis. Postoperatively he developed hypotension and tachycardia and began to hyperventilate. The arterial blood showed
>
> | Postop day 1 | pH | 7.48 | [HCO$_3^-$] | 12 |
> |---|---|---|---|---|
> | 1:15 P.M. | Pco$_2$ | 16.8 | BE | -8.2 |
> | | Po$_2$ | 39 | | |
>
> At first the hyperventilation was ascribed to hypoxemia, but when hyperventilation continued when PaO$_2$ was raised to 83 by changing from O$_2$ by nasal prongs to O$_2$ by face mask it was thought that the patient's respiration was being driven by fluid overload. Early the next morning inspiratory rales were heard:
>
> | Postop day 2 | pH | 7.54 | [HCO$_3^-$] | 17 |
> |---|---|---|---|---|
> | 2 A.M. | Pco$_2$ | 20 | | |
> | | Po$_2$ | 43 | | |
>
> The patient was placed on 12 L O$_2$/min by nasal catheter, but both tachypnea and hypoxemia persisted. Renal failure and dissemination of infection led to the patient's death on day 4.
>
> | Postop day 2 | pH | 7.51 | [HCO$_3^-$] | 21.5 |
> |---|---|---|---|---|
> | 8:30 P.M. | Pco$_2$ | 26.5 | | |
> | | Po$_2$ | 58 | | |
>
> Comment: Although the initial pH was 7.48, there was a moderate metabolic acidosis.* This was revealed both by the low [HCO$_3^-$] (much lower than expected in simple CO$_2$ depletion as seen by the position of point ㊴ in Fig. 40) and by the BE of -8.2. The low arterial Pco$_2$ was not, apparently, a compensatory phenomenon but was added to (superimposed upon) the metabolic acidosis. The patient therefore had a mixed acid-base disturbance consisting of two simultaneous primary disturbances, namely, [HCO$_3^-$] depletion and CO$_2$ depletion, ([HCO$_3^-$] = 12, Pco$_2$ = 16.8).
>
> However, as noted, hyperventilation continued even after arterial hypoxemia (potentiometric) was abolished by raising the Po$_2$ to 83 mm Hg. Whether oxygen

---

* This sentence brings out one of the major inadequacies of traditional terminology.

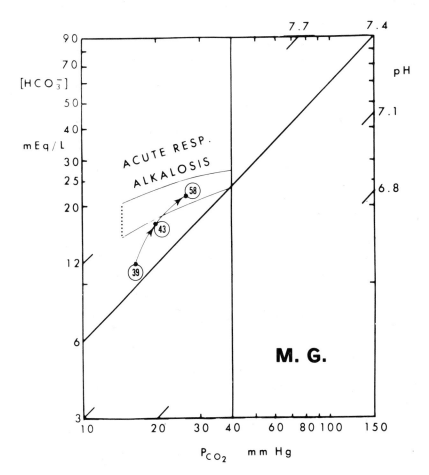

FIG. 40. The *oxygen* pressures (Po_2's) are shown in the circles as a means of visualizing four acid-base variables simultaneously. As the Po_2 rose so did [HCO_3^-], suggesting that lactic acidosis was lessening. However, hyperventilation persisted until the patient's death; the last point entered the expected significance band for acute respiratory alkalosis.[195]

    delivery to or the oxygenation of cerebral or other tissues was adequate was not revealed directly by the blood gas measurements. Lactic acid measurements might have been informative.

PRIMARY CO₂ RETENTION WITH AND WITHOUT ACIDEMIA

Clinical Description

    The signs and symptoms of CO_2 retention can seldom be separated from those of hypoxemia since these blood gas changes so often co-exist in patients with pulmonary disease. Carbon dioxide retention itself, acidemia of the blood and hypoxia "each has had its proponents as the principal cause for the cerebral symptoms."[129] Headache, drowsiness, confusion and slowly developing stupor are common as chronic hypoventilation becomes progressively more severe in chronic airway obstruction. The diagnosis of CO_2 retention is not difficult when cardiopulmonary failure is obvious; but when such patients have been gradually decompensating and then suddenly retain CO_2 because of an infection or ill-advised sedation, they may become comatose so rapidly that exogenous poisoning or other cause of coma may be erroneously suspected. Blood gas measurements make the diagnosis.

Chemical Description

In primary CO_2 retention the arterial P_{CO_2} is elevated as a result of hypoventilation not secondary to a base excess. Hence any $[HCO_3^-]$ rise can be considered secondary to the rise in P_{CO_2}.

An increased arterial P_{CO_2} only occurs with alveolar hypoventilation, i.e. ventilation which does not keep pace with CO_2 production. Acute CO_2 retention is associated not only with a rise in P_{CO_2} but with a fall in pH because bicarbonate concentration rises only slightly in a few minutes of hypoventilation. This "chemical rise" in $[HCO_3^-]$ can be seen to occur by the reaction depicted as follows:

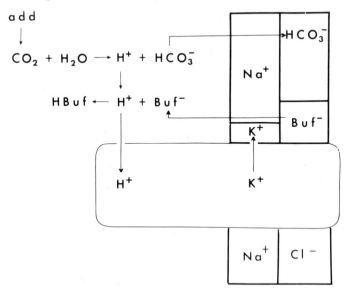

The same reactions occur as in CO_2 depletion except that their directions are reversed.

The greater proportional rise in P_{CO_2} than in $[HCO_3^-]$ causes the pH to fall:

$$\downarrow pH = 6.1 + \log \frac{[HCO_3^-]\uparrow}{.03\ P_{CO_2}\uparrow\uparrow}$$

As we have mentioned in Chapter 2, one reason $[HCO_3^-]$ in plasma rises only slightly in acute CO_2 retention is that, although it is chemically generated in the blood by the action of hemoglobin, it leaks out of the plasma into the ECF. At the same time K^+ moves from cells (especially skeletal muscle) [212,213] into extracellular fluid and into plasma. Such mobilization of potassium stores may contribute to potassium loss by the kidney, a loss which is characteristic of patients in respiratory failure and in experimental CO_2 retention.[214] A normal or even high serum $[K^+]$ may hide the fact that total body K stores have been depleted.

Means of Compensation

After a few hours to a few days of CO_2 retention, $[HCO_3^-]$ in plasma begins to increase above the slightly elevated level achieved by the chemical rise. This occurs, as we have discussed on page 66, by HCO_3^- "reabsorption" by the kidney. Typically, patients with chronically elevated P_{CO_2} do not increase this bicarbonate level enough to bring the pH completely back to normal, although there is a wide range of patient response.[215,216]

How much of the bicarbonate rise in a chronic pulmonary patient is "normal" or expected and how much can be attributed to another acid-base disturbance? To answer this question, Schwartz and his colleagues[7,195,196,197] have defined certain expected compensation limits or "significance bands" as possible aids in the differential diagnosis of acid-base disturbances. The usefulness of these bands is at present being considered in many clinical centers.

Clinical Causes

Hypoventilation can occur as a result of (a) insufficiency of central respiratory drive, (b) failure of the peripheral respiratory apparatus to respond to the drive, (c) failure of the chest bellows to provide adequate alveolar ventilation despite responding to the drive, (d) a changed respiratory drive-ventilatory response relationship. (The last is very common and the most complicated.)

(a) *Central nervous system depression or damage.* The unwise use of sedative drugs in unsuspected CO_2 retention can and often does convert a chronic steady state of hypercapnia to one of acute acidemia. (In addition, a rising P_{CO_2} always means a falling P_{O_2}; the reciprocity between P_{CO_2} and P_{O_2} should never be forgotten. The complaint of restlessness or insomnia in hospitalized patients is often a sign of hypoxemia which, with sedation, may rapidly become severe—especially, of course, in patients with pulmonary disease.)

Cerebral injury, although often increasing respiratory drive, may, if the medulla is damaged, greatly reduce it. So-called "idiopathic hypoventilation" and particularly an insensitivity to the stimulus of low oxygen are also important examples.

(b) *Failure of respiratory muscles to respond.* Spinal poliomyelitis, infectious polyneuritis and myasthenia gravis are examples in this category. Injectable antibiotics (kanamycin, polymyxin, neomycin and streptomycin) can all produce neuromuscular blockade and respiratory paralysis.

(c) *Severe acute airway obstruction or thoracic injury.* Acute foreign-body impaction, severe asthma, and copious secretions can so obstruct the airways that the respiratory muscles *cannot* prevent the Pco_2 from rising. In massive pneumo- or hemothorax, loss of rib-cage support from injury with paradoxical motion of the chest wall (flail chest), the respiratory muscles similarly cannot produce adequate alveolar ventilation.

(d) *Chronic obstructive lung disease.* In the commonest form of chronic CO_2 retention, the respiratory muscles *can but do not* provide adequate alveolar ventilation. Patients with chronic airway obstructive are "insensitive" to CO_2. This apparently results from a comparison (within the nervous system) between (1) the demands imposed on the respiratory apparatus (signaled by blood gas and afferent stimuli) and (2) the cost to the organism imposed by the work of breathing. The resulting level of ventilation in emphysematous patients has been aptly described as a "compromise adaptation,"[169] a concept previously developed by Richards and Barach[217] and by Riley.[14]

In 1932 Richards and Barach proposed an idea which few paid attention to: "With oxygen inhalation, oxygen absorption from the lungs into the arterial blood is facilitated, the arterial blood is normally saturated, or nearly so, with oxygen, while at the same time the pulmonary ventilation can and does diminish. But this latter state renders more difficult the elimination of CO_2. If, however, alveolar and blood CO_2 levels are increased, so that in a given expiration more CO_2 is eliminated, then equilibrium can be restored again. This is apparently the type of adaptation that takes place, both in cases of cardiac insufficiency and in those cases of pulmonary dysfunction that are relieved in high oxygen atmospheres."[217] Riley[14] put this slightly teleologic idea into quantitative terms; this, plus the fact that pulmonary mechanics was becoming recognized as relevant to medicine, gave the idea wide acceptance.

In brief, if the alveolar and arterial Fco_2 are allowed to rise, the % CO_2 in alveolar gas rises. Hence more CO_2 can be eliminated for a given volume of expired alveolar gas. Hence CO_2 output can be accomplished with less mechanical work.

Case 4. Respiratory acidosis complicated by metabolic acidosis to which were added metabolic and respiratory alkalosis.

R.C., a 39 year old known bronchitic, was admitted to the hospital because of increasing dyspnea of three weeks' duration. He had been on home O_2 for two years and had had ankle edema for one year. PE: T, 37°C; P, 88 irreg.; BP, 110/70; R, 60. Many premature ventricular contractions and rales to the midscapular level were noted. Serum electrolytes, except for Cl^- between 85 and 95 mEq/L, were normal and remained so. Initial blood gases on $2\frac{1}{2}$ L O_2 min:

Day 1 ①	pH	7.285	$[HCO_3^-]$	30
7:45 A.M.	P_{CO_2}	64	BE	+5
	P_{O_2}	50	% Sat	81%

The patient was not difficult to "oxygenate" and the arterial P_{O_2} never fell below 50. Four hours after admission, cardiac arrest occurred. A few minutes later the pertinent blood gases were

Day 1 ②	pH	7.26	$[HCO_3^-]$	33
12:45 P.M.	P_{CO_2}	77		
	P_{O_2}	53		

Cardiac resuscitation by epinephrine and tracheal intubation was successful. Thirty and 76 min later the blood gases were, respectively,

Day 1 ③	pH	7.09	$[HCO_3^-]$	19.5
1:00 P.M.	P_{CO_2}	66		
④	pH	7.275	$[HCO_3^-]$	18
1:45 P.M.	P_{CO_2}	40		

Despite $NaHCO_3$ the patient's dysrhythmia persisted. He could not trip the respirator himself and it therefore was cycled to maintain his ventilation. The next morning:

Day 2 ⑤	pH	7.65	$[HCO_3^-]$	41.5
8:00 A.M.	P_{CO_2}	37.5		

During the next two days, as the patient improved sufficiently so that adjustments in blood gas levels could be made, the measurements were, as shown by points ⑥ and ⑦ of Figure 41.

Comment: The patient was admitted in a state of chronic respiratory acidosis but this was decompensated because his pH was below 7.30. As seen in Figure 41, the first two points fall between the "acute" and "chronic" CO_2 retention areas. At points ③ and ④ he was being revived from cardiac arrest as a result of which his bicarbonate had fallen (probably by virtue of tissue hypoxia and lactic acid production). This metabolic acidosis was converted to a combined respiratory and metabolic alkalosis at point ⑤ when the pH rose to 7.65. The patient recovered without evidence that these acid-base disturbances had produced sequelae.

These four cases illustrate not only the classical acid-base disturbances and show how raw data (pH, P_{CO_2}, P_{O_2}, $[HCO_3^-]$, BE) can be (1) considered together and organized in the physician's mind, and related (2) to basic chemical reac-

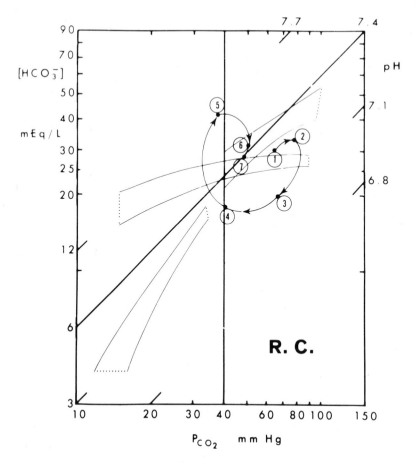

FIG. 41. Four primary disturbances occurring sequentially: CO_2 retention at ① and ②; [HCO_3^-] depletion with continued CO_2 retention at ③; pure HCO_3^- loss at ④; nearly pure HCO_3^- retention at ⑤; and proportional retention of both CO_2 and HCO_3^- at ⑥. The significance of the areas of expected deviation is not clear; they were not considered in the management of the patient.

tions and the Gamble electrolyte diagram,[218] (3) to the Henderson-Hasselbalch equation, and (4) to the classical diagram of Peters and Van Slyke.[28] Those familiar with the process of organizing such raw data may not need any of the last three "aids to understanding." Whether or not any or all of the four methods of presenting acid-base data will be supplanted in practical work by computer methods[219,220] is an open question. The following section illustrates how differential diagnosis can be achieved by simple clinical observation and chemical measurements.

THE DIFFERENTIAL DIAGNOSIS OF ACID–BASE DISTURBANCES

Formal knowledge of the four acid-base disturbances just described is necessary but not sufficient for their differential diagnosis. This is because, taken by themselves, many clinical and laboratory manifestations are *non-specific*. Thus asterixis (flapping tremor) occurs both in CO_2 retention and in the CO_2 depletion of liver disease. Similarly a low Pco_2 can occur in both diabetic and hepatic coma, and a high $[HCO_3^-]$ with both vomiting and chronic airway obstruction. Because stupor and coma are themselves such important although non-specific signs, we will consider how these and other such signs can be made useful by being correlated with blood gas measurements at the bedside.

The major clinical manifestations of acid-base disorders involve the central nervous system and are almost always reflected by changes in respiration. These manifestations are of great value in distinguishing between acid-base disorders. Nowhere has the ideal combination of bedside and laboratory observation in this field been better epitomized than in the words of Plum and Posner:

> "As a first step in appraising the breathing of patients with metabolically caused coma, increased or decreased respiratory efforts must be confirmed as truly reflecting hyperventilation or hypoventilation. Increased chest efforts do not spell hyperventilation if they merely overcome obstruction or pneumonitis, and, conversely, seemingly shallow breathing can fulfill the reduced metabolic needs of subjects in deep coma. Although careful clinical evaluation usually avoids these potential deceptions, the bedside observations are most helpful when anchored by direct determinations of the arterial blood pH, Pco_2, Po_2, and bicarbonate concentration."[129]

If a patient is comatose, the chances are about 7 in 10 that his brain is affected not by structural cerebral lesions but by "metabolic" changes— acid-base, hypoxic or toxic in nature. Before coma supervenes, disturbances of mentation, which nearly always precede coma as "warnings," should be looked for to head off the serious consequences of unrecognized acid-base disturbances. Presented with a patient who has already lapsed into coma, the physician must rapidly determine its specific cause.

The first question to be answered is whether hypoventilation is present or not because (1) most curable cases of coma are caused by exogenous poisons or endogenous toxic processes such as CO_2 retention and (2) hypo-ventilation is a threat to the cerebral oxygen supply. It is true that poisoning even by respiratory depressants like the barbiturates can produce transient hyperventilation and that both hypoglycemia and hypoxia can temporarily stimulate breathing. The resulting hyperpnea may be ineffective because breathing may be very rapid and shallow. Whether obvious or disguised, ventilatory failure calls for immediate action—assisted respiration via an adequate airway—and its presence or absence must be established either by observation or by determining the arterial P_{CO_2}, or both.

Hypoventilation

Hypoventilation in a comatose patient indicates either primary respiratory depression (primary CO_2 retention) or depression secondary to metabolic alkalosis (primary HCO_3^- retention).When caused by lung disease, which is usually obstructive, the arterial P_{CO_2} is likely to be between 50 and 100 mm Hg, and the pH below 7.30; when compensatory to HCO_3^- retention, the P_{CO_2} is seldom over 50 mm Hg and the pH is elevated (> 7.45). In both cases, $[HCO_3^-]$ is usually increased. If the primary pulmonary disease has been chronic, the bicarbonate will be comparable to that seen after severe vomiting, i.e. between 30 and 50 mEq/L. In both the base excess (BE) is elevated. Often in acute CO_2 retention the bicarbonate level is within normal limits; here the BE may be normal or "falsely" low. (See page 35.)

Both CO_2 and HCO_3^- retention can lower PaO_2. It will often be below 50 mm Hg in obstructive lung disease; it is less often lowered to this value in metabolic alkalosis.

The most generally useful way of distinguishing between the hypoventilation of respiratory acidosis and that of metabolic alkalosis is by an adequate assessment of the past and present pulmonary status. The arterial pH can clinch the differential diagnosis. Neither $[HCO_3^-]$ nor BE alone is adequate.

Hyperventilation

Hyperventilation in a stuporous or comatose patient is a sign of danger. It either means that the patient has a fixed acid excess and Kussmaul breathing or that his respiration is being driven by central or neurogenic influences. In the first case the pH is low in the arterial blood (less than 7.30 as a rule); in the second, the blood pH is likely to be 7.45 or above. (The P_{CO_2} taken by itself cannot be used to differentiate these two disorders; the $[HCO_3^-]$ and BE tend to be lower in metabolic acidosis than in respiratory alkalosis, but neither are as useful as the pH.)

Four important causes of fixed acid invasion are sufficient to produce coma: diabetes, uremia, poisoning (by drugs such as paraldehyde or materials like methanol or ethylene glycol), and abnormal lactic acid production. Urine and venous blood analyses are usually sufficient to diagnose diabetes and uremia. If these two are eliminated, poisoning should be suspected. During appropriate testing for evidence of intoxication, sodium bicarbonate should probably be used to treat the acidemia (if pH < 7.30 and $[HCO_3^-] < 20$ mEq/L).

Four important causes of primary CO_2 depletion can be associated with coma: salicylism, liver failure, certain types of pulmonary disease, and "post-traumatic pulmonary insufficiency" (see Cases 6 and 7). Salicylate poisoning causes mixed respiratory alkalosis and metabolic acidosis: the P_{CO_2} and the bicarbonate are both lowered, the P_{CO_2} usually more than $[HCO_3^-]$ so that pH is often elevated. A hyperpneic patient in stupor or coma with a normal or slightly elevated pH should be suspected of salicylism. Hepatic coma usually presents few diagnostic difficulties. Pulmonary edema and congestion occurring in an unconscious hyperventilating patient often lowers both the P_{CO_2} and the P_{O_2} in the arterial blood. Accurate diagnosis and management of comatose patients with extensive pulmonary involvement requires understanding both of acid-base and oxygen transport mechanisms. These are considered together in Chapter 6.

6

....

Blood Gas Interactions and
Oxygenation Disturbances

"In studying the interaction between oxygen and carbonic acid, it is of the first importance not to regard the change in one substance as cause and the change in the other as effect."   L. J. HENDERSON (1928)[3]

"Knowledge of intermediary metabolism and of enzyme systems has sufficiently advanced in recent years to prepare the ground for a study of the enzymic mechanisms which operate in the rate control of metabolic processes, and to give tentative answers to the question of how cells adapt their rates of energy supply to changing needs."
H. A. KREBS AND H. L. KORNBERG (1957)[120]

....

IN VIEW of the attempt in Chapter 4 to classify the physiologic mechanisms available for oxygenation in health, it would be logical to classify clinical hypoxic states in terms of the failure of these mechanisms. Because our knowledge of these states is largely limited to what has been learned from oxygen measurements in only one site—the arterial blood—the following classification is based more on expediency than on fundamental considerations.

We will try, however, to show how oxygenation disturbances are connected in fundamental ways to acid-base changes (particularly Pco_2 and pH levels), and to the relation between oxygen supply mechanisms and oxygen demands.

Oxygen and carbon dioxide pressures are normally reciprocal at many levels in the body. This reciprocity begins in the lungs where, generally, increases in one pressure occur with decreases in the other; it involves the chemoreceptor regulatory system; and it ends in the tissues where, because of enzymatic and other processes, Po_2-Pco_2 relationships are much more complex.

135

We will focus first on respiratory gas reciprocity in the lungs and arterial blood, only later attempting to deal with events occurring at the cellular level and with the clinically serious circumstances in which P_{O_2} and P_{CO_2} rise or fall together rather than reciprocally.

HYPOXEMIA WITH CO_2 RETENTION ($PaO_2\downarrow$, $PaCO_2\uparrow$)

In CO_2 retention during air breathing (asphyxia), alveolar P_{O_2} almost always falls more than P_{CO_2} rises. Arterial P_{O_2}, normally only about 10 mm Hg lower than alveolar, can be much lower than alveolar in pulmonary disease and can fall to levels associated with unconsciousness.

How much of an arterial P_{O_2} (PaO_2) lowering is to be expected solely as a result of CO_2 retention during air breathing at sea level? This can be estimated from the alveolar equation, a convenient approximation of which, for clinical purposes, is

$$P_{A_{O_2}}(\text{alveolar}) = 150 - \frac{6}{5} \times P_{A_{CO_2}}(\text{alveolar}).$$

Assuming an alveolar arterial P_{O_2} difference of 10, the expected PaO_2 would be, for a P_{CO_2} of 50 in alveolar gas and arterial blood (usually very similar):

$$PaO_2 = \left(150 - \frac{6}{5} \times 50\right) - 10 = 80.$$

In Figure 42 are shown arterial blood gas measurements obtained in 224 patients with $PaCO_2$ above 50 mm Hg. These are plotted in relation to a line showing the predicted reciprocal relationship between arterial tensions, assuming a normal R and a normal 10 mm Hg alveolar-arterial oxygen difference ($AaDo_2$). It can be seen that although the patients demonstrate P_{CO_2}-P_{O_2} reciprocity in their arterial blood, the P_{O_2} is always much lower than that predicted from R and normal $AaDo_2$ effects, a phenomenon partly explained in the previous chapter and further explained below.

A special reciprocity exists between P_{CO_2} and P_{O_2} as chemical stimuli. Respiration is normally driven by a high P_{CO_2} and (to a lesser extent under normal conditions) by a low P_{O_2}. How the absence of something (O_2) can stimulate is not yet clear. The stimulation is probably by a pH effect common to both O_2 and CO_2, for the chemical control of breathing occurs via changes in respiratory gas pressures in blood and/or pH changes near or inside chemoreceptor cells in the brain and in peripheral sites such as the carotid body and aortic arch receptors. When arterial P_{CO_2} and P_{O_2} are altered in a controlled systematic fashion, it can be shown that these chemical stimuli act synergis-

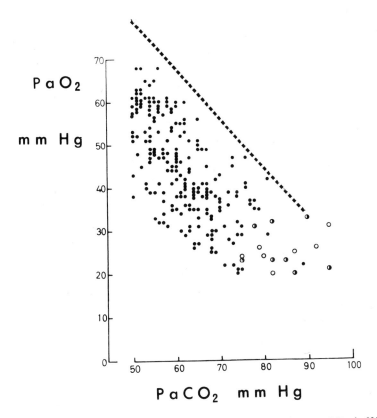

FIG. 42. An example of the reciprocal relationship between arterial P_{CO_2} and P_{O_2} in 224 CO_2-retaining patients breathing air. Closed circles: conscious patients. Half-open and open circles: semi-conscious and unconscious respectively (data of Refsum[216]). The dotted line is where all points would lie if they followed the alveolar equation, assuming alveolar-arterial P_{CO_2} equilibrium, an R.Q. $= 0.80$ and an $AaD_{O_2} = 10$ mm Hg.

The dotted line is steeper than 45° because of the R effect. R $= CO_2$ output/O_2 uptake and this is usually about 0.8; with asphyxia it may fall below 0.5. Consider prolonged breath-holding: O_2 will continue to be absorbed long after CO_2 transport from blood to gas phase has nearly ceased. This is because the P_{O_2} of mixed venous blood ($P\bar{v}O_2$) stays low longer relative to the gas phase than $P\bar{v}CO_2$ stays high as a result of the chemical properties of blood reflected by the different shapes of the O_2 and CO_2 dissociation curves. See Figure 43.

The R effect is exactly expressed by the alveolar equation:

$$P_{A_{O_2}} = F_IO_2(P_B - 47) - P_{A_{CO_2}}\left(F_IO_2 + \frac{1 - F_IO_2}{R}\right),$$

for sea level CO_2-free inspired air, where F_IO_2 is the inspired fraction of O_2 and P_B is the barometric pressure.

tically, that is, in a more than additive manner. Thus the ventilatory response to combined hypoxemia and hypercarbia is greater than would be predicted from the sum of the response to each stimulus acting separately.[184,221]

This synergistic stimulation may partially account for the acute distress of patients in pulmonary edema who often retain CO_2, develop acidemia and fail to oxygenate their arterial blood.[222] Because of laboratory evidence that low blood pH interferes with myocardial function, bicarbonate as well as oxygen therapy may be useful in managing patients with pulmonary edema with combined hypoxemia and CO_2 retention occurring acutely, although providing adequate pulmonary ventilation and raising blood pH in this fashion is probably a safer and more important form of treatment. (Whether an actual "metabolic acidosis" is always reflected by the finding[222] of a negative base excess or whether this finding results from bicarbonate migration from blood to extravascular fluid is not clear at present.)

Patients with hypoxemia plus CO_2 retention are generally those with chronic bronchitis and/or emphysema. As previously discussed, their increased work of breathing somehow changes their respiratory control system making them relatively insensitive to CO_2. As P_{CO_2} rises, P_{O_2} falls in the lungs. Some of the consequences of this reciprocity and of therapeutic interferences with it are clinically important even though poorly understood, involving both chemoreceptor regulation and metabolic changes at the tissue level (changes as fundamental as the demand for oxygen).[139,223]

When treating patients with chronic hypoxemia and CO_2 retention with high oxygen mixtures, P_{O_2}-P_{CO_2} reciprocity may be interfered with in two ways: not only may arterial P_{O_2} and P_{CO_2} rise and fall together but the oxygen drive (the "hypoxic stimulus") may become more potent than the CO_2 drive.

The importance of the hypoxic stimulus in human respiration was, of course, dramatized by the clinical observation that the administration of 100% O_2 can produce coma.[107,224,225] Many hypoxemic patients with chronic CO_2 retention are relatively insensitive to CO_2; their drive to maintain respiration seems to be chiefly via the carotid and aortic arch "low O_2" receptors. When 100% O_2 is given, these cease to be stimulated and ventilation falls. Under these conditions "the physician should be on his guard against deepening hypercapnia."[226]

It is a commonplace thesis in operating rooms and intensive care units that oxygen requirements take precedence over CO_2 homeostasis and that critically ill patients must be adequately oxygenated (if need be by mechanical ventilation) regardless of the (usually temporary) acid-base disturbances which may accompany therapy. (As we have seen in Case 4, even quite marked deviations from normal arterial blood acid-base values can be well tolerated.) Furthermore, the fear of "CO_2 narcosis," occurring in acutely ill CO_2-retaining patients, has, in the last few decades, been largely dissipated by better understanding of how to control* the arterial Po_2 with oxygen therapy.[227,228,229] In brief, the aim of O_2 therapy is to maintain—continuously—the arterial Po_2 between about 70 and 90 mm Hg, using whatever means are necessary to achieve this. Some patients will need only nasal O_2 at 1 L/min; others may need a tracheostomy and mechanical ventilation with high O_2 mixtures. If no more O_2 is given than that needed to bring arterial Po_2 to near normal levels, the rise in Pco_2 is usually minimal.

Patients with PaO_2 less than 30 mm Hg are, as we have mentioned, often unconscious or at least somnolent. Their survival is not strictly dependent on the level of PaO_2, *per se*, since, in a recent report, *more than half of 22 patients with PaO_2 from 7.5 to 20 mm Hg recovered.* An acute rise in Pco_2 to 80 mm or above, partly because of the intense acidity it can produce in blood and tissues, is associated with a poor prognosis, although, in general, "one cannot determine whether the hypercapnia and acidemia are dangerous in themselves or in combination with hypoxemia.[230]

In patients with chronic airway obstruction, the Pco_2 is chronically high and the Po_2 chronically low. If Pco_2 rises further, the arterial pH may well fall below 7.20, as in an acute exacerbation of chronic bronchitis. Granted that such patients require oxygenation. Do they also require correction of their arterial acidemia? It is generally believed, largely on the basis of laboratory experiments,[231–234] that "acidosis," or acidemia, or tissue acidity are associated with poor myocardial function and a tendency toward dysrhythmias and therefore that the arterial pH should be maintained within normal limits (7.35 to 7.45).

To extrapolate wisely from laboratory experiments lasting minutes to hours to the longer-term experimental situation brought about by the intensive care of acutely and chronically ill patients is a major medical problem today. Chronic disease usually brings about a chronic adjustment of many "normal levels" of measurable values to other levels. When (and particularly how fast) should the physician lower a high arterial Pco_2? When is returning a Pco_2 to "normal" contraindicated? The following case is typical of most physician's experiences in that it illustrates many problems but provides few answers.

Case 5. A 75 year old veteran with chronic bronchitis and emphysema had been followed for two years. During this time his $PaCO_2$ slowly rose from 40 to 54 where it remained for six months before he was admitted to the hospital. He had been on nasal oxygen at home at 2 L/min—administered by nasal prongs.

---

* How to *lose* control of arterial Po_2 has been clearly analyzed by Campbell.[227] If O_2 is administered intermittently to a CO_2-retainer (to try to keep the respiratory center "awake" and thus avoid CO_2 narcosis), dangerous hypoxemia can occur because, as air breathing is suddenly imposed, the alveolar Po_2 falls precipitously by the effect of removing a high O_2 source in the presence of a high alveolar Pco_2 which soon becomes higher.

Two weeks before admission his sputum became more copious and his dyspnea worse. An outpatient arterial blood analysis made just prior to admission revealed

Day 1   pH   7.31 P_{CO_2} 74 P_{O_2} 37
12:15 P.M.   (Patient still on 2 L/min O_2 via nasal prongs)

T, 96.8; P, 110; BP, 110/80; R, 34 (on admission). Patient was cyanotic and acutely dyspneic. Breath sounds consisted of wheezes. Liver down 4 cm; 2+ ankle edema. Hematocrit, 51; WBC, 13,000; K^+, 4.8; Cl^-, 96; "CO_2", 40.

During the first hour, assisted breathing via face mask, mouthpiece and IPPB was attempted but was ineffective because the patient struggled against the machine. The blood gases were:

Day 1   pH   7.22 P_{CO_2} 104 P_{O_2} 84
1:15 P.M.

A nasotracheal tube was passed but, although PaO_2 rose to 290 on 40% O_2, the P_{CO_2} rose to 118 and the pH fell to 7.18. The patient was therefore transferred to the Intensive Care Unit where a tracheostomy was performed and a volume respirator used. Five hours after admission the blood gases were

Day 1   pH   7.48 P_{CO_2} 49 P_{O_2} 92
5:15 P.M.

The patient looked better and was slowly improving with this assisted respiration, and with Keflin, aminophylline, digoxin and corticosteroid therapy. During the next 36 hours his P_{CO_2} was easily maintained between 39 and 41 (i.e. slightly higher than the normal Denver level of 34 to 38 but 15 mm Hg lower than his usual level). The P_{O_2} was easily maintained between 59 and 76.

At 10:15 P.M. a grand mal seizure occurred without warning. A lumbar puncture yielded clear fluid. It was not felt that the lowering of the P_{CO_2} had been too rapid. The impression was that a cerebral hemorrhage or infarction had occurred. Despite many therapeutic efforts, the seizures could not be controlled and the patient died 24 hours later. No autopsy was obtained.

Comment: This case illustrates the fact that the only way to be certain that assisted ventilation is effective in lowering $PaCO_2$ is to measure it. The measured rise in P_{CO_2} to 104 and fall in pH was felt to be justification for more vigorous ventilation. Despite the fact that arterial hypoxemia was under control, a tracheostomy was performed and ventilation given until the P_{CO_2} was lowered to near normal levels.

Two questions remain unanswered. If arterial P_{O_2} can be brought to normal but the pH remains below 7.20 in acute hypoxemia with CO_2 retention, is it mandatory to raise the arterial pH by lowering the P_{CO_2}, and if so, how fast and how much? Did this patient convulse because of therapy?[206,207] The answers to these questions will probably depend on more understanding of what is involved in O_2-CO_2 reciprocity and how this is determined by proton transfers in various tissues, especially the brain.

Acetazolamide (Diamox) might have stopped the convulsions of this patient by lowering CSF pH and/or increasing blood flow. In addition to its relatively slow action in promoting renal HCO_3^- excretion, and thus lowering brain HCO_3^- concentration, Diamox may increase cerebral blood flow directly by an important mechanism suggested by Severinghaus et al.[208] namely, by blocking the formation of molecular CO_2 from H^+ and HCO_3^-. This involves the novel assumption that bicarbonate (or H_2CO_3) rather than molecular CO_2 is the end product of decarboxylation in mitochondria—in other words, that the reaction

$$CO_2 + H_2O \overset{\text{Diamox}}{\underset{\text{block}}{\rightleftharpoons}} H_2CO_3 \rightleftharpoons H^+ + HCO_3^-$$

(in, say, vascular muscle cells) runs from right-to-left rather than in the usually assumed left-to-right direction. It is clear that Diamox would make H^+ "pile up" (in these cells) and lower pH if the direction were right to left, thus explaining the cerebral vasodilating action of Diamox by a pH effect.

Another explanation is that Diamox produces local hypoxia of blood vessels because O_2 release from hemoglobin might be impaired for two reasons: less CO_2 would enter red cells and that which did enter would produce fewer Bohr-Haldane protons (see Fig. 47). In the red cell, then:

$$\text{(from tissues) } CO_2 + H_2O \overset{\text{Diamox}}{\underset{\text{block}}{\longrightarrow}} H^+ + HCO_3^-$$
$$\downarrow$$
$$\text{(to tissues) } \quad O_2 + HHb \longleftarrow H^+ + O_2Hb^-.$$

This conclusion, however, is not supported by animal experiments[235] in which Diamox increased rather than decreased the P_{O_2} at the surface of the brain. How and especially under what conditions Diamox increases cerebral blood flow remains unclear.

HYPOXEMIA WITHOUT CO_2 RETENTION ($PaO_2\downarrow$, $PaCO_2$ NEAR NORMAL)

The chronic form of this type of hypoxemia occurs in congenital heart disease (with right-to-left shunts) and, more commonly, in pulmonary disease, especially in patients with diffuse pulmonary fibrosis and infiltration. Typically the $PaCO_2$ is normal despite an increased minute ventilation. This is explained by the statement that the dead space in such patients is very high,[236] although what accounts for this high dead space is obscure.[237] What drives ventilation in these patients is equally obscure; relieving their hypoxemia may scarcely lower their ventilation. It is widely held that abnormal reflexes from their stiff lungs are responsible.

In the numerous acute pulmonary disturbances which produce it, this type of hypoxemia is often insidious in onset since an arterial P_{CO_2} rise signaling respiratory failure is minimal or absent. Even minor pulmonary parenchymal disturbances (e.g. basal atelectasis) may lower arterial P_{O_2} significantly; unless this is recognized and treated the PaO_2 may become low and the P_{CO_2} be scarcely affected.

When is an arterial P_{O_2} *dangerously* low? When and how must it be treated? Answers to these questions depend largely on the type of pulmonary lesion which causes the arterial P_{O_2} to fall while scarcely affecting arterial P_{CO_2}.

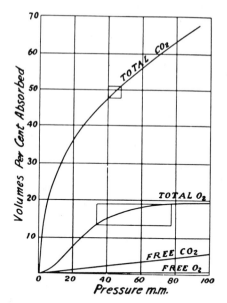

FIG. 43. CO_2 and O_2 dissociation curves of the blood of A. V. Bock plotted on the same coordinate system (from L. J. Henderson, 1928).[3] The arteriovenous Po_2 difference is much larger than the Pco_2 difference as shown by the greater horizontal width of the O_2 rectangle.

The arteriovenous O_2 *concentration* difference is also greater than that for CO_2 as shown by the greater vertical height of the O_2 rectangle. (This over-simplified figure makes A.V.B.'s R.Q. = 0.5; as Henderson pointed out, the correct value of 0.8 is obtained when the Haldane and Bohr effects and the greater magnitude of the former over the latter are taken into account. See Fig. 47.)

What type of lesion is usually responsible for such hypoxemia without CO_2 retention? We will deal with this question in three ways: chemically, pathophysiologically and with a clinical example.

1. Chemical Reasons for Arterial Hypoxemia without CO_2 Retention

This topic can be explained by considering, on an appropriate coordinate system, the shapes and slopes of CO_2 and O_2 dissociation curves. As shown in Figure 43, these curves differ in three major ways.

The CO_2 curve is higher than the O_2 curve. This reflects the fact that the quantity of CO_2 contained in blood (concentration) per unit of gas intensity (pressure) is much larger than in the case of O_2. Oxygen is concentrated in erythrocytes whereas CO_2 (mostly as HCO_3^-) is abundant in the extra- and intracellular fluids of both lungs and tissues. These differences contribute to the fact that CO_2 elimination is relatively ventilation dependent while the uptake of oxygen is more blood-flow dependent, a conclusion with obvious therapeutic significance.

The CO_2 curve is nearly straight, the O_2 curve sigmoid.   The major consequence of this difference is, as we have discussed in Chapter 4, that adequate
oxygen unloading pressure can be maintained as blood changes from arterial
to venous in the body tissues. The S-shaped curve of oxygen also partly
explains why ventilation-perfusion disturbances in the lungs lower arterial
Po_2 so much more than they raise the Pco_2.

The CO_2 curve is steeper than the O_2 curve.   The steepness of the CO_2
curve (already discussed in a different connection in Chapter 2) can only be
rationally compared to that for O_2 if the coordinates are the same for both
curves, that is, if the units of both quantity and intensity are made identical.*
When this is done, as in Figures 43 and 44, two explanations become evident.
(1) The relative steepness of the CO_2 curve explains why pure shunts have
so little effect on arterial Pco_2 (because a given quantity of CO_2 added to
alveolar capillary blood raises mixed arterial CO_2 tension so little). (2) The
relative flatness of the O_2 curve explains why hyperventilation of the lung or a
portion of it often cannot compensate for what is probably the commonest
type of pulmonary lesion causing arterial hypoxemia in clinical medicine—a
lesion causing *part* of the pulmonary artery blood to leave the lung with a Po_2
lower than in alveolar gas.

* It is no accident that the same coordinate system for the two gases was used in a modern
lung model,[238] showing how the relative steepness of the CO_2 and O_2 curves explains their
transport implications.

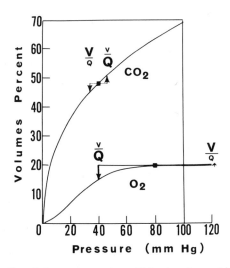

FIG. 44. Henderson's dissociation curves redrawn[239] to show how a high ventilation to perfusion ratio (high V/Q) has a greater effect in lowering CO_2 content than in raising O_2 content;
therefore, either a local or a general hyperventilation can compensate for a local CO_2 retention but cannot make up for the effect of a low V/Q on mixed arterial blood oxygenation.

2. Pathophysiologic Causes of Hypoxemia without CO₂ Retention

The three classical causes of arterial hypoxemia (general hypoventilation, shunt, diffusion defect) tend to lower PaO_2 more than they raise $PaCO_2$*; so does the usually mentioned fourth cause (a V/Q disturbance), as illustrated in Figure 44. In a given patient it is often impossible to sort out which one or combination of these four causes is present in the face of the over-riding tendency of the chemistry of blood to produce the same arterial gas changes with almost any localized pulmonary lesion.

The fact that many kinds of pulmonary lesions lower PaO_2 more than they raise $PaCO_2$ is, nevertheless, not sufficiently appreciated. It was recently said that "the idea that alveolar hypoventilation can exist without an elevation of $PaCO_2$ jars the pulmonary physiologist."[240] This statement should jar the clinical pulmonary physiologist who may think that the important distinction between general and local hypoventilation is widely enough understood in medicine. It is, however, gradually being appreciated that even in mild pulmonary disturbances (e.g. caused by anesthesia) "carbon dioxide tension is not a reliable index of adequate ventilation for the maintenance of normal arterial oxygen tension"[241] and that mechanically increasing *total* pulmonary ventilation postoperatively (for example) is not indicated for arterial hypoxemia. "Instead, the postoperative regimen calls for careful monitoring of arterial and venous blood gases, for the use of sufficient inspired oxygen to produce an arterial oxygen tension of 100 mm Hg, for intensive chest physiotherapy with coughing and deep breathing exercises, and for passive deep breathing every 30 minutes around the clock."[242]

Anesthetists have tended to explain hypoxemia by a single postulated lesion ("microatelectasis"). This is partly based on the finding that when sedated patients take a deep breath (or when a deep breath is imposed during anesthesia) the PaO_2 may rise. In addition to the over-riding effect of the chemical factors we have discussed, cardiac output and O_2 uptake changes (which can be considerable during anesthesia) can affect arterial oxygenation. As Prys-Roberts *et al.*[243] concluded, the hypoxemia of anesthesia "has a multiple aetiology."

3. A Clinical Example

A most important syndrome causing severe hypoxemia without CO_2 retention has been called "post traumatic pulmonary insufficiency."[172] Ashbaugh and his colleagues[244,245] have called the condition of such patients with

---

* General hypoventilation by the R effect, shunting because of the relative steepness of the CO_2 dissociation curve and diffusion disturbances by the greater permeance of CO_2 as compared with that of oxygen.

hypoxemia and stiff lungs the "adult respiratory distress syndrome." The syndrome not only occurs after non-thoracic trauma but from a variety of causes, e.g. drug overdosage, respiratory paralysis and aspiration pneumonia. The extreme respiratory distress associated with congestion and stiffness of the lungs, with profound hypoxemia and other blood gas changes, and often with terminal lactacidemia, tissue hypoxia and death can only be properly managed by monitoring arterial pH, Pco_2 and Po_2 (at least) and preferably $[HCO_3^-]$ or base excess as well. An example of the use of blood gas and other measurements follows:

Case 6. Adult respiratory distress syndrome.

S.V., a 52 year old white woman, was admitted to the hospital in coma following a barbiturate overdose. T, 37°C; P, 130; BP, 70/?. The lungs were clear but she was slightly cyanotic. An endotracheal tube was placed and dialysis begun. With an inspired Po_2 (P_IO_2) of 580 the blood gases were

Day 1	pH	7.235	$[HCO_3^-]$	18
12:05 A.M.	Pco_2	44	BE	−10
	Po_2	460		

The acidemia was soon controlled with bicarbonate and, at first, hypoxemia could be easily overcome with oxygen by T-tube since the initial P_IO_2-PaO_2 difference (IaD) was 580 − 460 = 120 mm Hg. However, with persistent hypotension and efforts to treat this with fluids, the PaO_2 rapidly dropped (see Fig. 45) and the patient began to exhibit tachypnea. A tracheostomy was done but the patient "fought" the respirator and became alkalemic and hypoxemic—a common early sign of the "respiratory distress syndrome" (RDS).

Day 2	pH	7.479
2:00 P.M.	Pco_2	33
	Po_2	57.5

Curare was used with good effect and P_IO_2 levels of 350 to 450 were instituted via a volume respirator along with CPPB. (Continuous positive pressure breathing has been found particularly important in combating hypoxemia in this type of patient. McIntyre et al.[246] have also found this true in "severe respiratory failure.") On the third day the patient was tried on room air but her arterial Po_2 fell to 37 mm Hg. Therefore O_2 was re-started and, with a P_IO_2 of 355, the blood gases were:

Day 3	pH	7.40
7:45 P.M.	Pco_2	42.5
	Po_2	93

Thus in three days the IaD had risen to 355 − 93 = 262—a most important feature of RDS. After an accidental disconnection occurred (on the fifth day) between tracheostomy tube and respirator, this IaD rose to over 300 temporarily as seen in Figure 45. From the seventh day on, however, this difference narrowed and the patient made a complete recovery.

Comments: This case demonstrates two of the four "phases"[172] of this syndrome. In the first low-flow phase, the patient was alkalemic just as so often occurs soon after major injury or acute hemorrhage.

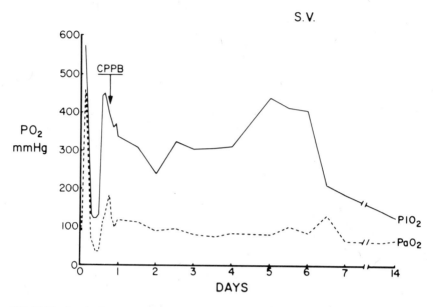

FIG. 45. The inspired oxygen pressure (P_IO_2) and the arterial oxygen pressure (PaO_2) in the respiratory distress syndrome. On the first day the PaO_2 dropped as low as 40 mm Hg and only rose to 180 on 100% O_2. The difference between inspired and arterial Po_2 (IaD) remained high for six days despite continuous positive pressure breathing (CPPB), but eventually returned to normal levels as oxygen transport across the lung improved. (From Ashbaugh *et al.*,[245] with permission.)

Respiratory distress ushered in the second phase in which *hypoxemia develops despite continuing hyperventilation*. Both PaO_2 and $PaCO_2$ are low (see Day 2). Paralysis of the patient for a few hours with curare eliminated the metabolic load imposed by the patient's struggles to breathe against rather than with a mechanical ventilator.

The third phase (progressively impeded pulmonary O_2 transport) was almost completely avoided by the use of continuous positive pressure breathing which Ashbaugh *et al.*[245] have found to be very helpful, probably because it seems "to reverse pulmonary edema and maintain alveolar patency" in the very stiff lungs of such patients.

The impeded O_2 transport in these patients is often called pulmonary "shunting." Strictly speaking, this term means that pure venous blood bypasses the lungs and enters the arterial stream producing hypoxemia by mixing with well-oxygenated capillary blood. Actually much of the hypoxemia usually results from V/Q disturbances or diffusion abnormalities which partially oxygenate venous blood. For this and other reasons, to calculate the shunt from the alveolar-arterial O_2 difference or the inspired-arterial difference (IaD) is not only inaccurate but conceptually misleading. See Figure 46.

Because of the physical chemistry of blood, arterial hypoxemia without CO_2 retention can result from almost any parenchymal pulmonary distur-

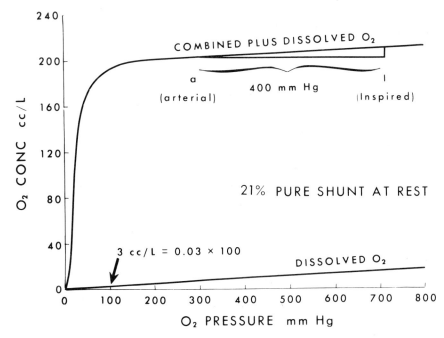

FIG. 46. When 100% O_2 is breathed at sea level and inspired P_{O_2} is increased to 760 − 47 = 713 mm Hg ("I" in the figure), any pulmonary capillary blood achieving equilibrium with alveolar P_{O_2} has its hemoglobin fully saturated. With even a small shunt, arterial P_{O_2} ("a" in the figure) will be 100 to 150 mm Hg less than inspired, because a slight lowering of arterial O_2 concentration by venous admixture results in a large P_{O_2} drop. This Ia difference increases as the shunt increases. The % shunt is roughly equal to Ia/20. More detailed shunt calculation requires knowledge of the alveolar-arterial P_{O_2} difference (Aa) and the arterio-mixed venous O_2 concentration difference (av) or O_2 uptake/cardiac output. The formula (for a > 150 mm Hg and normal hemoglobin conc.) is: % shunt = 0.03 × Aa/(0.03 × Aa + av). In the example Aa = 350 mm Hg and av = 40 cc/L so that % shunt = 21.

Large "shunts" are the hardest to calculate accurately. "If the amount of blood passing through the shunt is so great that the O_2 tension of arterial blood is less than 150 mm Hg, the chemically bound O_2-content of arterial blood must be calculated from the O_2-saturation read off a standard O_2 dissociation curve." This quotation is from Bartels *et al.*[247] who have given a complete statement of the difficulties of shunt calculations.

bance. To label such hypoxemia "shunting," carries with it the temptation to calculate the shunt. This is a procedure extremely liable to error, particularly in patients whose cardiac output is not measured and in anesthetized patients whose cardiac outputs (and probably their demands for oxygen) can be acutely raised and lowered simply by varying their arterial P_{CO_2}.[243] Even when cardiac output is measured, it is not always justified to *assume* an oxygen consumption (which is often done) because this can decrease with increasing levels of P_{CO_2}.[139,223] (An example of the difficulty of interpreting alveolar and arterial blood Po_2 levels on the basis of shunt calculations has recently been published.[248] The conclusion that increasing arterial Pco_2 reduced pulmonary shunting depended on the assumption that O_2 uptake was not affected by changing Pco_2 levels.)

In summary, because so many different pulmonary parenchymal lesions can produce the same low PaO_2 and normal $PaCO_2$, because current physiologic methods for distinguishing between these lesions are so difficult to apply clinically, and because calculations of "shunting" are so subject to error in serious illness, it is considered wise to be guided at least as much by clinical signs in the treatment of hypoxemic patients without CO_2 retention as by arterial blood gas values—especially indirectly derived values. It is probably true that any patient with an arterial Po_2 lower than 50 mm Hg should be treated with oxygen to bring this to between 70 and 90 mm Hg. In circumstances requiring positive respirator pressures greater than 50 cm H_2O or P_IO_2 levels greater than 200 mm Hg for many hours to days, an arterial Po_2 of 55 to 65 mm Hg may be all that can be achieved and, indeed, may be high enough to tide the patient over an acute episode. The primary consideration, as usual in treating a critical illness, is whether or not the patient has a reversible lesion justifying the use of heroic measures.

(Although these figures reflect experience in treating patients acclimatized to one mile above sea level, it is believed that the survival of sea-level patients being maintained at a PaO_2 of 55 to 65 mm Hg would be similar.)

HYPOXEMIA WITH CO_2 DEPLETION ($PaO_2\downarrow$, $PaCO_2\downarrow$)

Patients who spontaneously hyperventilate and yet do not properly oxygenate their arterial blood present many challenges. Not only are they often severely dyspneic from their inordinate respiratory "drive," but the question often arises today whether or not their arterial Po_2 is low enough to require oxygen therapy and their arterial pH high enough to cause concern. Patients with myocardial infarction, pulmonary embolism, shock and various forms of acute respiratory embarrassment are today being treated with oxygen in intensive care units for their hypoxemia. Because their hypoxemia is often associated with a *low* arterial Pco_2 and a high pH, they may also be treated for their alkalemia.

It is becoming clear that the proper intensive management of such patients is not limited by technical or mechanical ability but by lack of understanding of the factors that aid or impair such life-determining functions as cardiac output, myocardial rhythmicity and vasomotor regulation of blood flow distribution. Although there is evidence that hypoxemia, hypocapnia and high pH (separately or together) disturb these functions,[98,249,250] no unifying scheme has emerged which shows how these linked disturbances produce their effects. The state of knowledge in this field is indicated by a recent statement of Ayres and Grace[251]: "The frequency of alkalemia and hypoxemia in critically ill patients, . . . and its development from ventilation inappropriate to metabolic need have led us to the descriptive phrase 'inappropriate pulmonary ventilation in the critically ill' simply for emphasis and teaching."

The fact that arterial hypoxemia often follows myocardial infarction,[252–254] together with present-day ability to determine the arterial Po_2 easily, has stimulated many to study the effects of administering oxygen[255–257] in this condition. A recent editorial nicely balances "practical" reasons for O_2 therapy against proven efficacy as follows: "From a practical viewpoint the routine administration of oxygen to all patients with acute myocardial infarction is indicated" . . . but "final appraisal of the benefit of oxygen must be sought at the level of cellular physiology."[258] Since this is the level where our ignorance is greatest, such final appraisal is likely to take longer than is practical.

Acute myocardial infarction is caused by ischemia which not only (1) produces tissue hypoxia by depriving the heart of oxygen and substrates but also (2) simultaneously imposes end-product accumulation and tissue acidity. Separation of these two factors, even in the laboratory, is complicated by difficulty in controlling such factors as sympathoadrenal activity and the Bohr-Haldane effect in hemoglobin reactions.[259,260] Recent macromolecular research indicates that a low intracellular pH "by its effect on the Ca^{++} sensitivity of the heart's contractile proteins, may play a major role in causing the precipitous decline in the contractile force of the ischemic heart,"[261] but under most clinical conditions the effects of arterial hypoxemia, myocardial hypoxia, and abnormal pH of blood and myocardium cannot be separated. Under these conditions, treatment must be highly empirical.

A low arterial Po_2, a high pH (early) and low pH (late) have been found to occur in circulatory failure from hemorrhagic and cardiogenic shock.[262–264] The arterial Pco_2 in hemorrhagic shock tends to be between 30 and 40 mm Hg; the total ventilation is normal or increased. The arterial hypoxemia has been said to result from "shunting" of venous blood through the lung, but, as two recent reports emphasize, this is not pure shunting in the sense that pure mixed venous blood enters the arterial stream but a far more complex oxygenation disturbance thought to result from atelectasis, pulmonary edema and pneumonia.[265,266] Whatever the pulmonary mechanisms, both hypovolemic and septic shock are often associated with an arterial Po_2 as low as 40 mm Hg. Thus, cellular hypoxia is probably partly caused by arterial hypoxemia in addition to being caused by low O_2 delivery to the tissues as a result of ischemia. However, because of the complexity of the shock problem and the lack of a unifying scheme for understanding disturbances at the cellular level, it is still true (as written in 1943) that "the status of oxygen therapy [in shock] is not yet fully defined although there has been much investigation of this subject over many years."[110]

Severe tissue hypoxia almost always produces lactic acid and the "insult of acidosis" (where?) is therefore often superimposed on poor oxygenation. Is local lactic acid production always an insult?* May it be a mechanism for locally counteracting the respiratory alkalemia of hyperventilating patients?

In most of the patients just discussed a state of spontaneous hyperventilation exists. What are the implications of the resulting non-reciprocal blood gas pattern, i.e. of a low PaO_2 and a low $PaCO_2$? This will be discussed in connection with pulmonary ventilation, the rate of O_2 uptake and the behavior of hemoglobin.

1. Ventilatory Effects of Po_2-Pco_2 Interaction

It took many years for the full clinical realization of the *mechanical* implications of a low PaO_2 and a *high* $PaCO_2$—namely, that more CO_2 can be eliminated with less breathing work and hence with more O_2 left over to be utilized by tissues other than the respiratory muscles. (How the control centers match the need for CO_2 removal against the O_2 cost of removing it is still unknown.)

It took even longer for laboratory work to show the regulatory implications of the reciprocity of Pco_2 and Po_2—namely, that they interact *as stimuli* and that, therefore, they cannot be fully understood separately. In 1919 Haldane et al.[268] showed that the effect of diminishing oxygen percentage is to lower the threshold for CO_2. Since that time most attempts to quantitate one stimulus have either neglected the other or improperly controlled it. Thus "physiologists, having attempted the separation of the stimuli, have been strangely loath to put them together again until quite recently."[221]

It is now known that there is no sharp Po_2 threshold below which oxygen tension must fall to stimulate respiration, nor one above which high O_2 becomes inhibitory. In fact, oxygen at high pressure (more than one atmosphere, OHP) may be *stimulating* to the CNS by increasing local acidity as a result of blocking the Haldane effect. Thus, "hyperoxygenation clearly has the capacity both to stimulate respiration and to depress it, and these effects may occur together."[269]

---

* The fact that the prognosis of patients with post-traumatic pulmonary insufficiency is very poor if the lactic acid level in *blood* is greater than 10 mEq/L ($\sim$90 mg%) does not prove that the lactic acid itself is harmful to *brain* or to *myocardial function*. Long ago Haldane wrote: "It seems to me a gross mistake to treat the supposed acidosis of anoxaemia due to gas poisoning, 'shock,' and other conditions by the administration of alkalis. The body is calling for both oxygen and acid."[267]

If the effect of CO_2 change is removed (by strict maintenance of isocapnia) a smooth continuous ventilation-response-to-low-O_2 curve can be demonstrated under a wide variety of conditions.[184,270] Such holding-one-variable-constant experiments do not imply that O_2 effects can only be understood if P_{CO_2} is kept at one constant level—far from it, because it was the experimental approach of controlling both gases at several levels which led to the understanding of the synergism between CO_2 and O_2 stimuli as first clearly shown by Nielson and Smith.[271] What they do imply is that the "natural" situation of CO_2-O_2 reciprocity can only be understood by *putting together* separately studied stimulus-response curves. A clear example of the way synthesis can be achieved by proper use of analysis is seen in Cunningham *et al.* who conclude their paper with: "the response to hypoxia accompanied by hypercapnia—the 'natural' situation—is part of an inverse relation that continues far above the levels of P_{O_2} that could have been encountered before the discovery of oxygen and the possibility of artificial enrichment with oxygen of the atmosphere breathed. The hypoxic stimulus is an important component of the intense stimulation caused by asphyxia, and the response to it under natural occurring conditions is much greater than has commonly been supposed."[221]

What about the unnatural condition of low P_{O_2} and low P_{CO_2}? This not only occurs in the diseases we have been discussing in this section, but this hypoxemia-hypocapnia combination can be readily imposed by overventilating patients with impaired pulmonary O_2 transport, especially when they are anesthetized or artificially paralyzed.

2. *Effect of P_{CO_2} Levels on O_2 Uptake*

Recent work[223,272] has shown that, in anesthetized and paralyzed dogs (conditions similar to those imposed in intensive care units), considerable changes in metabolic rates (O_2 uptake, $\dot{V}_{O_2}$) occur when P_{CO_2} levels are changed. Even in conscious man, if his ventilation is controlled, the whole body O_2 uptake is inversely related to the levels of alveolar P_{CO_2}.[139] Of specific interest here is the fact that alkalemia from any cause seems to produce an *increased* metabolic rate in isolated tissue.[273] On the other hand, alkalemia caused by hypocapnia can produce a marked *decrease* in cardiac output under certain conditions (e.g. in anesthetized man[243]).

It seems as if the O_2 supply system (mainly the ventilatory and circulatory reserve mechanisms of Chapter 4) can be reciprocally related to the O_2 demands of the body. Perhaps, under certain conditions of clinical importance, the same arterial blood gas changes are associated with opposite responses in the supply and demand systems.

To keep some of these relationships clearly in mind, they have been collected in Table 11. The symmetry of the scheme, and the fact that O_2 and CO_2 probably affect ventilation by proton (pH) changes in chemoreceptors, suggest that energy metabolism as reflected by O_2 uptake may similarly be linked to changes in H^+ activity at the cellular level. This latter possibility is supported by recent findings which "affirmed the importance of pH in setting $\dot{V}_{O_2}$ of anesthetized paralyzed dogs."[223]

TABLE 11. *A brief summary of how respiratory gas pressures reciprocally can affect pulmonary and cellular activity*[139,221,223,243]

Ventilatory Stimulation $\uparrow Pco_2, \downarrow Po_2$	Ventilatory Depression $\downarrow Pco_2, \uparrow Po_2$
—————————————H+——————————————	
O_2 Uptake Stimulation $\downarrow Pco_2, ?\uparrow Po_2$	O_2 Uptake Depression $\uparrow Pco_2, ?\downarrow Po_2$

H^+ transfer and pH change probably form the common link between these blood gas stimuli.

"The effects of intracellular $[H^+]$ are common to chemoreceptor cells and to other tissue cells in general. The relatively small effect on $\dot{V}_{O_2}$ is probably a non-specialized 'primitive residual' which may be more important to unicellular organisms than to us (I'm guessing). The much more potent effect on chemoreceptor cells is realized in their output response and its magnification thereby is precisely the result of specialized function. That is, all cells are chemoreceptors! Some are more so than others. The mechanism by which the chemoreceptor increases its impulse train while decreasing its energy demand in the face of a high $[H^+]$, I will be happy to leave to the membrane scholars."[318]

3. Hemoglobin Behavior as a Model for Enzyme Action

As an attempt to bring some unity to these diverse considerations, we conclude this section by discussing the reciprocity of CO_2 and O_2 via proton transfer at the level of the red cell (and possibly other cells). Reasons for suspecting that too much CO_2 and too little O_2 act as a ventilatory stimulus (and perhaps the reverse combination as an O_2 uptake depressant) are implicit in the fundamental discovery[274] that regulatory mechanisms depend on changes in the shapes of enzymes produced when the occupancy of one binding site affects the binding properties of another. The hemoglobin molecule, although not an enzyme, is the *prototype of this enzyme behavior*[275] because, as O_2 is released from red cells, the molecule's affinity for protons increases, a "Bohr-Haldane" proton enters it (by H^+ transfer), changes its conformation, and reduces its O_2 affinity (Bohr effect). Simultaneously the Haldane effect (of even greater quantitative significance) occurs—also via proton transfer: the Bohr-Haldane H^+ is transferred from the Pco_2-bicarbonate system thus generating HCO_3^- and preventing a Pco_2 rise which would otherwise occur from CO_2 production. This HCO_3^- generation is depicted by the uppermost arrow and the uppermost star in Figure 47.*

The formulation of this figure is simply an illustration of the proton transfer implied by the equation given by Peters and Van Slyke:[28]

Blood in tissues:

$$K_2HbO_2 - O_2 + 0.5\,H_2CO_3 = K_{1.5}H_{0.5}Hb + 0.5\,KHCO_3.$$

We call the proton involved a "Bohr-Haldane proton" because, although Wyman, who demonstrated[277] that the Bohr effect is due to a discharge of protons resulting from the binding of O_2 by Hb, called them "Bohr protons," they are also intimately concerned with the C—D—H or Haldane effect.

---

* This figure does not attempt to include carbamino compounds whose contribution is still under investigation.[276]

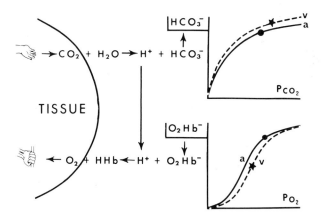

FIG. 47. How the Haldane and Bohr effects are linked by proton transfer. When CO_2 is released from tissue and reacts with water in the blood (first reaction), H^+ is produced and is transferred (in the red cell) to O_2Hb^-. This fairly strong base becomes much stronger as it unloads O_2 to the tissue (second reaction). By the first reaction HCO_3^- is generated, the arterial blood becomes venous and absorbs more CO_2 (as HCO_3^-) (upper star) with only a slight increase in Pco_2 (Haldane effect). By the second reaction $[O_2Hb^-]$ decreases as arterial blood becomes venous, but so does pH. This pH drop shifts the O_2 dissociation curve to the right (lower star) so the blood unloads O_2 with only a slight decrease in Po_2 (Bohr effect). (With permission.[164])

It is clear that as protons become bound by the base Hb^-, more HCO_3^- is generated than if the blood had stayed oxygenated and that, as a result, less of the incoming CO_2 can remain free to exert a pressure. At the final equilibrium in venous plasma (star on venous CO_2 dissociation curve of Fig. 47), the P_{CO_2} and pH are thus scarcely different than in arterial plasma.

This intricate mechanism is obviously affected by changes in P_{CO_2} and P_{O_2} which are normally reciprocal. What do non-reciprocal changes do to this mechanism and are the effects similar to what may be occurring in cellular enzyme systems? A partial answer lies in the study of hyperbaric oxygen effects: it is known that high oxygen pressure may be toxic by *blocking* this type of proton-transfer with consequent increased acidity (either in blood or within tissue cells).

The highly practical subject of oxygen poisoning is beyond the scope of this text, but it is touched on here because the "high-tissue P_{CO_2}" explanation for oxygen toxicity is one which allows for direct manipulation of intracellular environments by therapy, namely, by changing P_{CO_2} levels and hence the pH near and inside of cells. It has been amply demonstrated[278,279] in animals and man that slight elevations (by 10 to 20 mm Hg) of P_{CO_2} enhance the convulsive effects of OHP (oxygen pressures greater than the atmosphere)—a demonstration that supports the classical notion that high oxygen, by interfering with the Haldane effect and hence with CO_2 removal from the tissues by the blood, produces toxic effects by increasing tissue acidity. However, it is not widely appreciated that higher P_{CO_2} levels (80 to 150 mm Hg, not uncommonly seen in chronic pulmonary patients on oxygen therapy) *protect* animals against the convulsive effects of OHP perhaps by influencing "the manner in which oxygen affects the enzyme systems of brain tissue."[280] (Are CO_2-retaining patients on home oxygen therapy similarly protected? Are hypocapneic patients being treated with high oxygen mixtures unprotected?)

The Haldane effect "accounts for about half the change in the blood CO_2 that occurs in the respiratory cycle,"[281] but this effect and interferences with it are not to be thought of as being, in themselves, responsible for more than a fraction of the normal CO_2 and O_2 transport through body tissues or the failure, in disease, of this transport. However, the effect is very well understood and may serve, being a prototype for similar effects in many enzyme systems, as a model for explaining the remarkable fact that an increased P_{CO_2} can be, simultaneously, a ventilatory stimulant and an O_2 uptake depressant, while a decreased P_{CO_2} can be a ventilatory depressant and an O_2 uptake stimulant.

Case 7. Following an auto accident, S.D., a 43 year old woman, was admitted to the hospital in shock with a right hemopneumothorax, abdominal hemorrhage and a ruptured jejunum and bladder. After surgical repair and while on 40% O_2 administered by T-tube attached to endotracheal tube, the ABG were

Day 2	pH	7.20	[HCO$_3^-$]	14.8
10:30 A.M.	Pco_2	39	BE	-12
	Po_2	55		

Respiratory distress syndrome was feared and a tracheostomy was done. On a volume respirator with a P$_1$O$_2$ of 410, the arterial blood was oxygenated but at the cost of alkalemia due to low Pco_2:

Day 2	pH	7.52	[HCO$_3^-$]	24
10:35 P.M.	Pco_2	30		
	Po_2	58		

Pulmonary rales, alveolar infiltrates demonstrated by X ray, persistent anxiety and a tendency to "fight" the volume respirator characterized the course. The pH remained above 7.50 and the Pco_2 in the low 30's. Despite maintaining the PaO$_2$ above 60, the patient's respiratory drive continued. Hypotension, renal failure and pseudomonas infection contributed to the patient's death nine days after the accident.

Comment: Typical case of post-traumatic pulmonary insufficiency with unexplained respiratory drive. It is impossible from the data available in this case (or of the vast majority of such cases) to answer the question, "Were the arterial blood gases properly controlled to provide the optimal environment for the cells of the heart, brain and kidney?" Until a unifying scheme is evolved for dealing with this type of question, it will probably remain unanswered.

TISSUE HYPOXIA WITH OR WITHOUT OTHER DISTURBANCES ($\downarrow P_{T_{O_2}}$)

Clinical examples of this type of oxygenation disturbance, documented with measurements of low tissue Po_2, are virtually non-existent. In most patients with respiratory failure, congestive heart failure, shock or severe injury, the evidence of tissue hypoxia is indirect or inferential only.

If mixed venous Po_2 (in blood sampled from the pulmonary artery) is lower than about 30 mm Hg, the presumption is that the tissues in general are poorly oxygenated, but this information tells nothing about the state of the brain or myocardium in particular. If the arterial lactate is above 2.0 mEq/L, the presumption that anaerobic sources of energy are being called on, but hyperventilation, as we will see, can also raise the blood lactic acid concentration.

Low tissue Po_2 and lactacidemia often occur together in clinical oxygenation disturbances. Because hypoxic tissue acidity is receiving much attention in neurophysiology[282] and because of the special vulnerability of the brain to hypoxia, we will conclude this chapter by considering how acid-base and oxygenation processes over which we have some control (by changing patients' ventilation) can affect cerebral function and how they are regulated via the pH in blood and extravascular fluids.

Of the four mechanisms discussed in Chapter 4, the first or circulatory reserve is probably the major one the organism uses to protect the brain against hypoxia. Adequate cerebral blood flow is partly assured by the generalized systemic response to stimulation of the vasomotor centers in the pons and medulla. Any slowing of blood to these centers is believed to produce local medullary elevations of Pco_2 (or fall in pH by lactic acid production) and a consequent powerful "CNS ischemic response"[127] consisting of widespread systemic sympathetic vasoconstriction.* Even when this response fails, auto-regulation of cerebral blood flow prevents (in the absence of cerebral injury) ischemia over a wide range of blood pressure changes. The mean systemic blood pressure must fall to 70 to 80 mm Hg before blood flow to the brain decreases significantly.[283]

How do changes in ventilation contribute to or defend against cerebral oxygen lack? When CO_2 is breathed or retained by the body, cerebral blood flow increases because of cerebral vasodilatation. Conversely, hyperventilation, by lowering arterial Pco_2, tends to constrict cerebral vessels and to reduce blood flow to the brain. Hyperoxemia (high arterial Po_2) produces cerebral vasoconstriction while hypoxemia dilates cerebral vessels. With this background let us summarize how hypoventilation and hyperventilation affect the brain, when CO_2 and O_2 pressures and pH are all changing simultaneously.

1. Hypoventilation

Increased arterial Pco_2 usually produces contrasting effects on cerebral and systemic arterioles. It dilates cerebral vessels and usually constricts those in the general systemic circulation. Because of the resulting combined increase in cerebral blood flow and systemic blood pressure, the brain is well guarded against tissue hypoxia even though the arterial Po_2 may fall with hypoventilation. When, however, acute airway obstruction is suddenly superimposed on chronic CO_2 retention (as in acute exacerbation of chronic bronchitis), unconsciousness may occur. Its cause in detail is unknown. The arterial Po_2 is usually about 25 mm Hg in such patients (as seen in Fig. 42), but it is doubtful that a sharp unconsciousness threshold of 25 mm Hg of Po_2 in arterial blood

---

* "The circulatory effects of excess CO_2 are dual and opposite: (1) by a direct stimulant action on the vasomotor center (augmented by reflex stimulation from the carotid and aortic bodies), it increases cardiac rate and force of contraction and narrows vessels with a sympathetic vasoconstrictor innervation; (2) by a direct action on arterioles, it dilates them. In conscious man, the first action predominates, and blood pressure and heart rate increase; when the vasomotor center is unable to respond (because of deep anesthesia, brain damage, severe ischemia or anoxia) or is disconnected from the peripheral parts of the sympathetic nervous system (because of spinal anesthesia or blocking drugs), direct vasodilation is the only or dominant effect and blood pressure falls."[107]

is a very useful clinical predictor of unconsciousness in these or in acutely ill patients in general.

In recent years, considerable understanding of the importance of the permeance of CO_2 in cerebral tissue has emerged. The regulation of the pH of the CSF and brain tissue and how this is related to changes in the blood is under active investigation,[204,282,284,285] but is a highly controversial and confusing subject. Posner and associates[10,317] have offered a valuable explanation of the mental state of patients with acid-base disturbances which has brought together many diverse observations—chemical, physiologic and clinical. They showed that, in respiratory acidemia, the fall in CSF pH parallels the fall in pH in the blood resulting from CO_2 retention. On the other hand, in metabolic acidemia the CSF pH may be normal or even elevated despite a low blood pH. How can blood pH and CSF pH move in the same direction in the first disturbance and in the opposite direction in the second? As seen in Figure 48, these results are largely explained by the ease with which CO_2 crosses the blood-brain barrier in comparison to bicarbonate and dissociated fixed acids. It is important to note the direction CO_2 takes as well. The permeance of CO_2 as it moves between blood and brain tissue satisfactorily explains why patients with respiratory acidemia tend to be more obtunded and more often in coma than patients with metabolic acidemia, even if both have an identically low pH in their arterial blood.

In general, states of hypoventilation, even unsteady states in which ventilation is suddenly lowered, are temporarily well tolerated by the brain and by the body tissues because both a rising P_{CO_2} and a falling P_{O_2} bring about protective responses at many levels in the organism. The unifying factor is probably the pH in extravascular fluids; convincing evidence for this has been incorporated in a recent neurophysiologic model.[286] This can be summarized as follows: cerebral vasomotion is controlled by the pH environment of arterioles, i.e. by the ratio of slowly changing extravascular $[HCO_3^-]$ to that of rapidly changeable P_{CO_2}. Since tissue P_{O_2} is so largely controlled by local blood flow, it is clear that the permeance of the CO_2 molecule is a key factor in cerebral oxygenation and that the brain's oxygen supply is critically affected by blood P_{CO_2} changes.

2. Hyperventilation

If both arterial P_{CO_2} and P_{O_2} are low, how do the cerebral vessels respond? When patients breathe 6% O_2 in N_2 and the P_{CO_2} falls (as a result of the hypoxemia-induced hyperventilation), no change in cerebral blood flow occurs until PaO_2 falls below 30 mm Hg.[287] When PaO_2 falls below 30 mm Hg, cerebral blood flow increases 56% despite an average decrease in P_{CO_2} to 34.1 mm Hg. Despite this increase in CBF, the O_2 uptake of the brain stays

RESPIRATORY ACIDOSIS

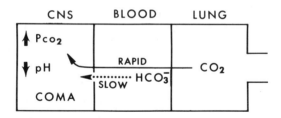

METABOLIC ACIDOSIS

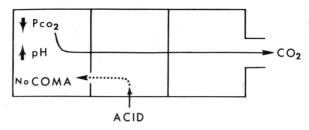

FIG. 48. When CO_2 is acutely retained in the lung it rapidly moves through the body down Pco_2 gradients. It can cross the blood-brain barrier much faster than HCO_3^-. In the upper figure the general direction of CO_2 movement is depicted from lung to CNS, raising the latter's Pco_2 and lowering its pH, thus, perhaps, causing coma. (Of course, the Pco_2 of the brain could rise simply because cerebrally produced free CO_2 cannot escape, the cerebral blood being high in Pco_2 along with all other blood.)

In diabetic coma, however, (for example) respiration is stimulated by a *low* pH caused by a primary invasion of acid into the blood. This acid may (relatively slowly) enter the CNS. If, however, CO_2 is washed out of the brain rapidly enough, the pH may be high in the CNS yet low in the blood.[10]

constant. Therefore, progressive potentiometric hypoxemia (by dilating cerebral vessels) can *override* the vasoconstrictive effect of the accompanying hypocapnia in the presence of intact circulatory and respiratory regulating mechanisms.

Certain types of hyperventilation, because they cause cerebral ischemia, are thought to cause cerebral hypoxia as well. When severe hypocapnia is induced by voluntary or imposed hyperventilation while the subject is breathing room air, it may be that cerebral hypoxia is more likely to occur than when the hypocapnia occurs as a response to arterial hypoxemia. Plum and Posner[203] have shown that imposed hyperventilation with air leads to a rise in CSF lactate, a rise which exceeds the rise in the blood. When hypoxemia (arterial Po_2 between 45 and 65 mm Hg) was added to hyperventilation, the CSF lactate increase was greater than with hyperventilation alone. Plum and

Posner reasoned that cerebral hypoxia "caused only a portion of the lactate rise produced by hypocapnia," but felt that "unfortunately, the reasoning is only tentative"[203] because the validity of the excess lactate concept of Huckabee[288] is in question.

Because lactic acid production has long been known to be associated with tissue hypoxia and because hypocapnia can cause cerebral vasoconstriction, it is not unreasonable to believe that hyperventilation with air can cause cerebral hypoxia. The meaning of increased lactate in cerebral tissue is, however, far from clear. It may be the result of increased anaerobic metabolism or (a very different thing?) may result from alkalemia or alkalinity of the environment of cells.[289]

Some parts of the brain are more susceptible than others to the effects of ischemia. This, and the fact that hyperventilation is being consciously and unconsciously used in the treatment of medical, neurosurgical and many other conditions, suggest that the effect of a low blood Pco_2 on the brain should be considered under two headings, iatrogenic and spontaneous.

Iatrogenic hyperventilation for diagnosing and treating cerebral lesions. Hyperventilation is utilized routinely as part of electroencephalography. Patients suspected of cerebral irritability are asked to hyperventilate; this procedure often brings out evidence on the EEG of focal or diffuse abnormality.

Neurosurgeons have used hyperventilation for years for cerebral edema in attempting to reduce brain volume; the exposed brain can be made to shrink in minutes. In cerebral infarction (where local autoregulation of vessels is impaired), there is some evidence that lowering arterial Pco_2 may decrease the extent of tissue injury.[290,291] The hypocapnia may increase the perfusion pressure in the small vessels in the area of the injury, or may, by correcting local tissue acidity, normalize the permeability of these vessels. However, since hyperventilation increases CSF lactate and causes EEG slowing[292] which improves when 100% O_2 is breathed,[293] it remains an open question whether or not hyperventilation has an overall beneficial effect on cerebral infarction.

Iatrogenic hyperventilation for bringing the arterial Pco_2 to "normal." Patients with chronically high Pco_2 (if greater than 50 the patient almost certainly has obstructive lung disease) are often mechanically ventilated to relieve this hypercapnia and the commonly present low pH. Too rapid a lowering of arterial Pco_2 has produced convulsions in such patients.[206,294] (See page 140.)

Spontaneous hyperventilation caused by known abnormalities. Apprehension, arterial acidemia and hypoxemia commonly cause hyperventilation. These respiratory drives are understood primarily because they can often be partly or wholly removed by treatment following which hyperventilation ceases.

Spontaneous hyperventilation caused by poorly understood abnormalities. Hyperventilation may persist in the absence of known respiratory drives (e.g. in hepatic coma, cerebral injury and congestive heart failure). In critically ill patients with acute pulmonary lesions (pneumonia, atelectasis, embolism, acute respiratory distress with non-compliant lungs), such persistent hyperventilation induces "increasing therapeutic and diagnostic frustration"[156] on the part of the physician. In these patients, arterial blood Po_2 and oxygen content may be normal or above normal; the Pco_2 is usually low. Our ignorance of what drives the respiration of such patients (who frequently thrash about in bed and "fight" the respirator) makes appropriate action extremely difficult. The assumption is often made that tissue hypoxia somewhere is responsible.

Until better information is available, perhaps it may be advantageous to bring together in a logical fashion four variables which affect cerebral function. These are pH, Pco_2, Po_2 and lactic acid. Figure 49 indicates a by now familiar framework into which future studies of cerebral acid-base and oxygenation processes may or may not be fitted. It may help interpret the following four quotations from the recent literature:

"CSF lactate finds its origin in the nervous tissue and not in the blood."[295]

"In hypoxia there are optimal conditions for an outflux of lactic acid into the CSF."[289]

It may be that "CSF pH is purposefully regulated by . . . diffusion of lactic (and pyruvic) acid from tissue cells."[204]

"Phosphorylation of glucose and fructose-1-P" provides a mechanism which "would explain the increase in lactate and pyruvate in moderate hyperventilation and the decrease in hypercapnia."[289]

3.   *Lactic Acid, Tissue Hypoxia and Hyperventilation*

It has been known for many years that heavy exercise, ascent to high altitude and injury severe enough to produce "shock" are associated with increased lactate levels in the blood. The common factor explaining this increase has been generally considered to be tissue hypoxia. It is not so generally appreciated that each of these conditions is associated with hyperventilation.

Hyperventilation leads to high lactate levels especially if it is imposed mechanically.[296] Mechanical (assisted) ventilation often leads to a low cardiac output and thus to a great excess of ventilation over blood flow in the lung and a respiratory alkalosis. Among the many explanations[202,288,297–300] for the lactate rise accompanying hyperventilation, the effect of *alkalinity* (high pH in blood and/or tissues) on glycolytic enzyme systems seems to be the common factor.[296,301]

The converse relation between pH and lactate levels has also been established; that is, hypercapnia and acidity (in blood and tissues) lowers lactate levels.[302] It is of great interest that β-adrenergic blockade has the same effect

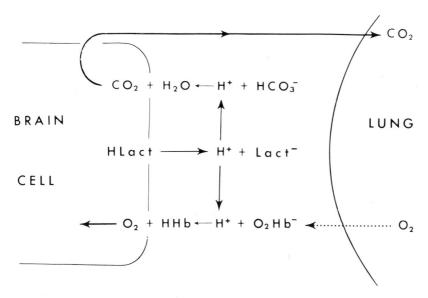

FIG. 49. Outline of the major acid-base reactions in the cerebral fluids accessible to analysis (CSF and blood) as they probably occur during hyperventilation with air. Very little O_2 is added to blood in the lungs (dotted arrow) in comparison to the CO_2 removed from lungs and hence from brain cells (upper arrow). The central role of the proton H^+ is evident. It is possible that H^+ production increases O_2 unloading from the blood (by the Bohr effect) thus counteracting one effect of hypocapnia.

as hypercapnia[303] because this strongly implies that catecholamine calorigenesis is inhibited by both influences. As we have discussed previously, pH changes in, or in the environment of, metabolizing cells may be the common factor in the regulation of oxygenation.

Until the many effects of hyperventilation on lactate levels are better understood, blood lactic acid measurements as quantitative indicators of tissue hypoxia should be used with caution.

Case 8. J.L., a 24 year old girl, was admitted to a mountain community hospital 12-23-68 after being found unconscious at 3:30 P.M. in an apartment (at 8,000 feet) with a faulty furnace. She was treated for CO poisoning with tracheostomy and O_2, and for the (presumed) ⟨acidosis⟩ with $NaHCO_3$. She remained in semi-coma for four days during which time her lungs became partly consolidated; she was then transferred to a medical center. T, 39; P, 120; BP, 118/60; R, 30. She had persistent decerebrate rigidity, positive Babinski's signs, persistent tachypnea and rhonchi, and 3 + pitting edema. X rays showed bilateral diffuse alveolar infiltrates.

Forty centimeters of H_2O pressure were needed to achieve and maintain a tidal volume of 650 cc. The low compliance of the lungs was treated with continuous positive pressure breathing (CPPB). After 15 minutes on a controlled volume respirator, the arterial blood gas measurements were:

5:30 P.M.	pH	7.60	[HCO_3^-]	28.5
(without added	Pco_2	30		
dead space)	Po_2	55	(Inspired Po_2 345 mm Hg)	

7:45 P.M.	pH	7.54	[HCO_3^-]	34
(with added	Pco_2	40		
dead space)	Po_2	48	(Inspired Po_2 395 mm Hg)	

The marked impediment to pulmonary O_2 transport is evident in the first two measurements of inspired to arterial Po_2 difference (IaD). They were IaD = 290, IaD = 247. Arterial pH kept between 7.44 and 7.54 on Day 1.

Day 2. Valium used for extensor spasms. IaD was 134 on two occasions; pH was held between 7.44 and 7.47.

Day 3. Subcutaneous crepitation of neck and upper chest was noted, probably caused by high airway pressures imposed by the mechanical respirator. It caused no difficulty. IaD gradually fell to 125 on continuous positive pressure breathing with volume control. Valium was increased. "Becoming less rigid."

Day 4. "Chest x ray clearing nicely." "Expiratory hold can probably be reduced."

Day 5. IaD (9:00 A.M.), 131; IaD (4:00 P.M.), 117.

Day 6. More alert but at 7:15 P.M. sudden variations occurred in vital signs: BP varied from 128/82 to 160/110; P from 60 to 140; R from 16 to 30. Later BP was 116/70, P 88, R 16. Impression: acute airway obstruction from excessive secretions. The secretions were successfully removed. IaD 87. Still on respirator.

Day 7. IaD 56 (i.e. nearly normal for I_1O_2 = 174).

Day 35. Complete neurologic examination. The patient was discharged on day 39 greatly improved but with impaired memory, moderate spastic paraparesis and with some apraxia and ataxia. She made a complete recovery within three months of discharge.

Comment: The cerebral hypoxia caused by CO poisoning damaged the brain, but with treatment the patient eventually completely recovered. Was the hypoxia made worse initially by giving HCO_3^-, shifting the O_2 Hb curve to the left and further interfering with O_2 unloading from hemoglobin? The IaD fell daily after CPPB was instituted; there is evidence that this type of controlled breathing improves pulmonary O_2 transport (see Case 6).

Sudden variations in vital signs, in our experience with cases of combined cerebral and pulmonary insult, usually result from pulmonary rather than cerebral complications; adequate suctioning or adjustment of the arterial blood gases usually stabilizes these signs.

Prevention of pH increase was only possible with the help of arterial blood gas measurements; the pH is presumed to be an important variable to maintain near normal levels. In actual fact we do not know this to be true in cases of cerebral hypoxia, either in general or with specific types of cerebral lesions.

Conclusions

....

PRACTICAL ARTERIAL BLOOD GAS CONSIDERATIONS

IN A FIELD as rapidly developing as blood gas interpretation, definite rules are difficult to formulate. Some general considerations preface the present attempt.

The first question is whether or not to order a blood gas determination.

> "The suspicion that a blood gas estimation may be abnormal does not, in itself, constitute an indication for the investigation. For example, the patient with diabetic pre-coma who does not have superimposed chest problems is unlikely to be better off for having his pH and Base Excess determined. These will tell little more than will careful consideration of the results of the blood sugar, electrolytes and ketones, and as such will contribute little, if anything, to the patient's management. A second example might be the patient coming out of the operating room having had, say, a prolonged laparotomy not associated with pulmonary problems. Frequently such patients have a low arterial Po_2 when they are recovering from the anesthetic. However, a blood gas estimation is not the prime requisite here. What the patient really needs is medical attention to ensure that he is conscious, breathing adequately (a Respirometer may be useful here), to be encouraged to cough and to be given I.P.P.B. therapy to prevent or overcome possible small areas of atelectasis complicating the anesthetic. Only *after* such practical considerations have been taken into account should a blood gas estimation be made if there is any suspicion of respiratory insufficiency."*

The second question involves when to order an analysis and the query: "Is this the ideal time and do I know enough about the state of the patient to interpret the data?"

> "The results of blood gas estimations must be available and noted 'within the hour' of the blood having been drawn. Gas tensions and pH can sometimes change rapidly, depending on the patient's condition and the pathology involved, and if it is not sufficiently important to the clinician to know the results until some hours have passed, the chances are that the investigation was not warranted in the first place."

* These quotations are from written contributions of my associates, Doctors Philip Corsello, David Kelble and Val Lindquist.

"Is the patient in a steady state? On I.P.P.B.? Receiving O_2? How much? Has he just been suctioned? Is he about to be suctioned? It is of the first importance to know these things before the arterial blood is drawn."

The third question is, "How necessary are blood gas data in this case?"

"At the present time, arterial blood gas estimation, although simple and quick to perform, remains a specialized procedure. The fact that the facilities are becoming more readily available is to the advantage of both patients and doctors. However, increasing availability of any investigation can easily lead to its being used as a 'routine' procedure. Alas, when an investigation falls into this category it tends to lose a number of intrinsic values:

"1. Time eventually obscures the real motivation for having the investigation done in the first place, and questions that one should have asked oneself before tend to be asked after the results are at hand. Forethought and clinical acumen are thereby lost.

"2. It is sometimes forgotten that the patient's peace of mind takes precedence over that of his doctor's. The implication that an investigation is 'routine' and 'readily available' engenders the risk that it will be used on occasions primarily to allay the doctor's anxiety. The doctor may sleep more peacefully for knowing that the Po_2 is normal—but will the patient?

"3. The distinction between the words 'routine' and 'mundane' is frequently blurred. Both carry the hallmark of disinterest. It is not unknown for the results of routine investigations to be lost to both sight and mind simply because by nature they are considered to be mundane. The plasma aldosterone level may be more easily remembered than the hematocrit, say, when the latter is done as a matter of form on a patient who would not otherwise appear to have a blood disorder.

"4. Arterial puncture is a relatively simple procedure. Although the risks and complications (e.g. hematoma, arterial thrombosis) are minimal, one should not overlook the fact that, although completely painless to the doctor, it may not be so for the patient.

"5. In clinical medicine it is all too easy to lose sight of the economics involved in patient care, and over-zealous 'routine analysis' serves only to inflate medical costs. Although it is obvious that in practice no patient should be deprived of any procedure simply for financial reasons, it is worth remembering that in the end someone has to pay. Currently the cost of arterial blood gas analysis in this Center is about $30."

The following are four situations which the author believes usually warrant an arterial puncture:

Unexplained tachypnea, dyspnea or disability for exercise suggests the need for arterial blood analysis—especially in patients whose pulmonary and cardiac function, by more usual criteria, is not sufficiently abnormal to explain the symptoms.

Unexplained restlessness and anxiety in hospitalized bed patients all too frequently turn out later to result from hypoxemia—especially in elderly sedated patients not on O_2 therapy. Cyanosis is an unreliable sign.

Unexplained drowsiness or confusion should raise the suspicion of CO_2 retention, especially in patients on O_2 therapy. It is probably even more difficult to diagnose hypercapnia than hypoxemia from clinical signs alone.

Before elective major thoracic surgery, arterial blood analyses are often advisable. (An adequate chest history and spirometry may be equally helpful.)

Before prolonged O_2 therapy is started, arterial blood gas measurements are needed (1) to determine if O_2 is indicated, (2) to evaluate CO_2 retention and the possibility that it will become worse if O_2 therapy is given, and (3) as a base line to judge the efficacy of O_2 in improving blood oxygenation. Without such measurements, it may be very difficult to predict, after O_2 therapy is started, how low the PaO_2 would fall when O_2 therapy is discontinued. It may also be dangerous (see footnote, page 139). On the other hand, O_2 should not be withheld from a cyanotic, dyspneic patient if the arterial blood cannot be analyzed immediately.

RULES FOR BLOOD GAS INTERPRETATION

1. Obtain arterial blood for analysis.
2. Consider pH, Pco_2, $[HCO_3^-]$ and Po_2 in relation to normal sea-level values (7.40 ± 0.04, 40 ± 4, 24 ± 2, 85 ± 5). The normal PaO_2 is age-dependent (Fig. 50).
3. Ask, "Are the measurements believable?" (Consider their technical reliability.) "Internally consistent?" (Do they obey the Henderson-Hasselbalch equation? Plotting the three variables must yield only *one* point on the graph of Fig. 51.)

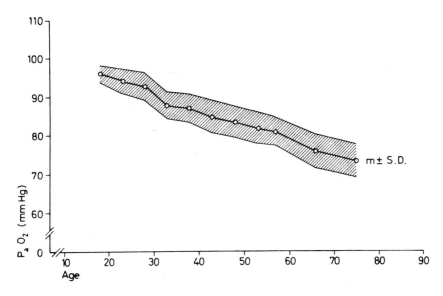

FIG. 50. The normal arterial Po_2 obtained in 152 supine subjects plotted in relation to age.[304]

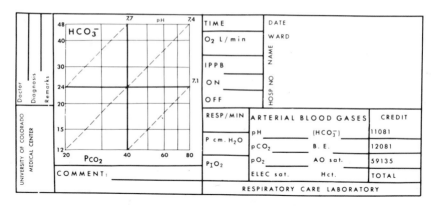

FIG. 51. A laboratory report slip for recording clinical blood gas data.

4. From knowledge of pulmonary, renal, gastrointestinal and other history and present status decide, on clinical grounds, which blood gas deviations are primary and which secondary. For example:

    a. In chronic obstructive lung disease, a high $[HCO_3^-]$ or positive BE would probably be a secondary change (but not with HCO_3^- therapy or duodenal obstruction and vomiting).

    b. In renal failure, a low $[HCO_3^-]$ would probably be a primary and a low Pco_2 a secondary change.

    c. In marked hyperventilation, the exact opposite can be the case: a low $[HCO_3^-]$ can be secondary to a primary lowering of Pco_2 (see pp. 72–74).

5. How much of the disturbance is physician induced—e.g. by diuretics, fluid therapy, antibiotics?

6. If there is a question that the blood may not be arterial, compare the values with simultaneously drawn venous blood (no tourniquet).

7. During assisted breathing, when any significant change in vital signs occurs, especially in cardiac rhythm, or change in level of anxiety or consciousness, an arterial blood gas analysis will often reveal the need for a new setting of the respirator.

8. Interpret blood gas data only with knowledge of the state of the patient and his therapy at the moment of arterial sampling.

9. Indications for therapy are primarily clinical; the following examples only show the order of magnitude of changes suggesting the need for therapy.

    a. pH < 7.25; pH > 7.55

    b. $Pco_2 > 50$ if pH < 7.30

    c. $[HCO_3^-] < 20$ or BE < -6 if pH < 7.30

    d. $Po_2 < 55$ if recent

    e. Clear evidence of tissue hypoxia

CONCLUDING REMARKS

After two decades of study of the hemoglobin molecule and how it links CO_2 and O_2 transport, a research chemist wrote: "It is now possible to see, in a general way, how the α and β-chains move in relation to one another as oxygen enters and leaves and the Bohr protons leave and enter, the whole structure behaving like a miniature reciprocating engine which mimics us, at a molecular level, as we breathe. . . . Until now those who have been trying to understand the mechanism of the hemoglobin molecule by taking off and putting on pieces and looking for what happens are still but as children playing with a watch."[275]

As respiratory physiology advances, it may be possible to see how oxygen transport is reciprocally linked to CO_2 transport in diseased tissues (probably via proton and electron transfers), but up to now our understanding of clinical oxygenation disturbances is very limited. However, if, in addition to taking off blood and analyzing it and putting on respiratory equipment to restore blood gas values toward normal, we look for what happens with enough physiologic insight, acid-base and blood gas chemistry will play its proper role in clinical medicine.

Appendix I

....

ON CLINICAL ACID-BASE TERMINOLOGY

ACID-BASE physiology is inherently a slightly confusing subject, but clinical acid-base terminology makes it very confusing. One inherent source of confusion is the unfortunate fact that the three most important classical acid-base measurements have magnitudes which are inversely related to the severity of the disturbances they reflect. Thus:

Arterial pH is inversely related to the severity of acidemia

Arterial P_{CO_2} is inversely related to the level of alveolar ventilation

Plasma $[HCO_3^-]$ is inversely related to the quantity of fixed acid excess

For reasons discussed elsewhere,[31,73-77] it is better to understand these inverse relationships than to try to change the fundamental units of the measurements (e.g. from pH to $[H^+]$).

It is also better to understand how certain ambiguous terms arose and to avoid being confused by them than to attempt to ban these from the language because of the practical impossibility of success in such an attempt.

The basis for the understanding of clinical acid-base disorders are the measurements obtainable in patients. As has been pointed out,[305] two languages have grown up which were intended to bring out the meaning of such measurements—a physiologic language which describes processes and a laboratory language which substitutes words for numbers. The following table demonstrates the contrast between the definiteness of measurements and the vagueness of the words used in both these languages to describe a typical acid-base disturbance.

Ambiguity in the meaning of words, a major curse of specialization, arises both because words often outlive the concepts they originally stood for and because the same word is used by different experts to mean different things, i.e. abused. The most abused and hence confusing official (non-slang) word in the acid-base field is probably "acidosis." Originally coined in 1906[306] to refer specifically to the intoxication by β-hydroxybutyric acid in diabetes, it has

TABLE A1. *Terms used to describe initial and recovery data on a patient with simple metabolic acidosis*

	Initial values	Values during recovery
Blood pH	7.178	7.475
Plasma pCO_2	18.0 mm Hg	25.0 mm Hg
Buffer base	27.0 mEq/L	40.0 mEq/L
Base excess	−21.0 mEq/L	−8.0 mEq/L
Plasma $[HCO_3^-]$	6.5 mEq/L	17.8 mEq/L
Physiologic language	(simple) metabolic acidosis with partial compensation	(?) "over-compensated" metabolic acidosis (?) respiratory alkalosis (?) mixed metabolic acidosis and respiratory alkalosis
Laboratory language	(?) primary metabolic acidosis and secondary respiratory alkalosis (?) mixed disturbance with primary metabolic acidosis and secondary respiratory alkalosis	(?) mixed disturbance with primary metabolic acidosis and primary respiratory alkalosis (?) primary respiratory alkalosis

From R. W. Winters,[205] with permission.

stood for a wide variety of chemical, physiologic, and clinical states ever since. Even today, when an exquisitely sensitive instrument is painstakingly used to measure the pH inside cells from which the bicarbonate concentration might be calculated, the even more sensitive instrument of language is so abused that one cannot tell whether "intracellular acidosis" means that the pH or the $[HCO_3^-]$ is low in the cells.*

The most confusing clinical slang term used in acid-base physiology is "CO_2" unmodified and unqualified. Since 1917 when Van Slyke and Cullen[307] developed the "CO_2-capacity" as an index inversely related to the severity of acid invasion, the term "CO_2" has been used for CO_2-combining power, serum CO_2 content or plasma $[HCO_3^-]$. These last three are measures of base *quantity* and entirely different from the CO_2 *intensity* factor (the P_{CO_2}) which is under

* I suggest that the former be designated as cytacidity (or, better, by giving the pH value) and the latter simply by stating the intracellular $[HCO_3^-]$ numerically.

the control of the respiratory system. Thus for 50 years the use of "CO_2" has made fuzzy the fundamental distinction between a non-respiratory disturbance (in which CO_2 content is abnormal) and a respiratory disturbance (in which CO_2 tension is abnormal).

Although "usage and not the reformer is the arbiter of words,"[102] this short glossary of especially confusing words has been compiled in the hope that some of their ambiguities may be at least lessened.

Outmoded or ambiguous usage	*Recommended usage*

Acid

An anion, e.g. Cl^-, $SO_4^=$.	A proton donor or H^+ donor.

Acid-base balance

A state in which the concentrations of Na^+, Cl^- or other electrolytes are in certain ratios, or in which pH is normal or an ill-defined subject concerned with proton transfer reactions.	The difference between the input and output of acid or base with respect to a defined system over a defined interval.

Acidemia

A state in which blood pH is below 7.36.

Acidosis

The finding or presumption of any of the following alone or in combination in any body fluid: low pH, low $[HCO_3^-]$, high Pco_2, low standard bicarbonate, low buffer base, or negative base excess.	An abnormal body process tending to lower pH but in which pH may not be low at all stages of the process; should be made definite (by adjectives such as respiratory, non-respiratory, lactic, etc.) or avoided.

Alkalemia

A state in which the blood pH is above 7.44.

Alkalosis

The finding or presumption of any of the following, alone or in combination in any body fluid: high pH, high $[HCO_3^-]$, low Pco_2, high standard bicarbonate, high buffer base or positive base excess.	An abnormal body process tending to raise pH, but in which pH may not be high at all stages of the process; should be made definite (by adjectives such as respiratory, non-respiratory, hypokalemic, etc.) or avoided.

Base

A cation, e.g. Na^+.	A proton-acceptor or H^+-binder.

Buffer (verb)

To minimize changes in $[H^+]$, pH, Pco_2 or Po_2.

To minimize changes in pH.

Compensated

Refers to a process by which the *blood* pH is returned to 7.4, the implication being that this is optimal in all circumstances.

To be avoided unless defined.

Content (of gases)

Generally means "concentration" in gas volumes STPD per unit volume of liquid.

Hypercapnia

A blood $Pco_2 > 44$ mm Hg (does not mean that CO_2 content is high).

Hypocapnia

A blood $Pco_2 < 36$ mm Hg (does not mean that the CO_2 content is low).

Metabolic

In acid-base jargon, either means "caused by gain or loss of any acid but carbonic" or a change in $[HCO_3^-]$ or BE not "accounted for" by a Pco_2 change.

Applicable to any biochemical process involving energy exchange. (Carbonic acid is the chief acid product of aerobic metabolism.)

Non-respiratory

Caused by gain or loss of any acid but carbonic.

Tension (of gases)

Partial pressure.

Appendix II

....

ON QUANTITY AND INTENSITY FACTORS

ALTHOUGH the significance of the difference between "intensive" and "extensive" variables has long been recognized in thermodynamics, it is not generally appreciated that a similar difference is of practical importance in acid-base chemistry, physiology and medicine.

1. *In Chemistry*

The most direct statement of the importance of this distinction is that made by W. Mansfield Clark:

> "Among the numerous developments of the theory [of dissociation] announced by Arrhenius in 1887 none is of more general practical importance than the resolution of 'acidity' into two components—the concentration of hydrogen ions, and the quantity of acid capable of furnishing this ionized hydrogen . . . thus the expression log $1/[H^+]$ is a linear function of the *intensity* factor of energy change and in this sense it can be called an index to acid intensity . . . we call 'normality' in its older sense the *quantity* factor of 'acidity.' "[31]

In 1908 L. J. Henderson, in the paper[25] which laid the foundation of acid-base physiology, wrote that "acidity of even very low intensity" does not develop in the living animal and "alkalinity of any real intensity is impossible" despite large changes in the quantities of acid and base in the body. Furthermore he clearly pointed out that the contrast between intensity and quantity factors is to be found not in their exact quantification, but in the fact that they differ by so many orders of magnitude as to be different in quality. On pages 433 and 448 he wrote:

> "It cannot be too strongly emphasized that the inevitable slight inaccuracies of calculation and determination of the equilibria here involved are as nothing compared with the difference between the enormous variations in amounts of base and acid, and the almost infinitesimal variations in hydrogen and hydroxyl ionization . . . In such a subject great precision is not demanded, because by the very nature of physiological processes variations and adaptability must always in certain respects prevent precision of definition. Accordingly, unless the present considerations shall prove to be in some

173

manner fundamentally incorrect or fallacious, the description of the physiological regulation of neutrality here presented must be a true description of the general nature of the process, a frame within which all details of the process may be pictured. It is, in that case, open to quantitative changes, corrections, and amplifications, but qualitatively it must be correct."*

Six years later Barcroft[103] was already trying to prevent "acidosis" from being thought of as an intensity factor: "Acidosis . . . will signify the appearance of acids (exclusive of CO_2), abnormal in kind or perhaps only in quantity in the blood. . . . But by the term acidosis I will signify nothing concerning the final 'reaction' of the blood . . ." In 1925 Barcroft[308] was using a special word—"acidaemia"—for a low blood pH, an excellent word which, however, did not gain official recognition until very recently.[309]

In 1916 Hasselbalch[40] dramatized the need for separating the concept of acid accumulation and that of a disturbed acid-base equilibrium by pointing out that unless a new term for the latter were devised it would not only be possible to have " 'Acidose' keine 'Acidose' " but even the opposite, i.e. acidosis with alkalosis. He strongly advised that "acidosis" be used to indicate the *accumulation* of acid in the blood and that a new term involving [H+] be invented to indicate the acid-base equilibrium or ⟨balance⟩, (Gleichgewichts).

Van Slyke[310] has consistently used "acidosis" in his writings to indicate a change in the quantity factor; in 1921 he clearly separated its meaning from that of a pH change. Michaelis[311] (whose classic work on the hydrogen ion was translated into English in 1926) called the quantity factor the "titratable acidity" and the intensity factor the "actual acidity." Bronsted, after several paragraphs dealing with this distinction, wrote (p. 244):

> "Although this is a purely formal question, it ought not to be dismissed as immaterial. The possibilities for development in a theoretical field are often dependent to a considerable extent upon suitable definitions and logical conceptions, whereas natural progress can be hampered or led astray if the definitions are lacking in clarity or cogency."[79]

The distinction between the quantity of material reacting and the intensity of chemical reaction is quite general. Wyman,[277] in describing the linked functions of large molecules, clearly shows the formal connection between (1) the fractional saturation of hemoglobin with oxygen at a given oxygen pressure, (2) the dissociation of protons at a given hydrogen ion activity and (3) the oxidation of substances containing several oxidizable groups as a function

---

* This extraordinary way for a master of quantitative chemistry to end a paper reveals the breadth of Henderson's mind and the mastery he knew he had of his subject. He, at least, never became infected by that medieval disease Clark[31] called "the pursuit of 'vanishing particulars' " which in this century often takes the form of the precise quantitation of the irrelevant.

of the voltage of the system—in all cases symbolizing the quantity factor by large X and the intensity factor by small x.

Most recently Frisell[82] has emphasized the difference between "actual acidity" (Michaelis' term) as measured by pH and the buffer capacity which is "determined by the total quantities of salt and acid."

2. In Physiology and Clinical Medicine

Strictly speaking a chemical quantity should be in moles (or equivalents or milliequivalents) and a chemical intensity in units of the chemical potential, e.g. calories per mole, because the product of a quantity and an intensity factor should have the units of energy.* Hence a concentration, expressed in milliequivalents per liter, is not a pure example of either a quantity or an intensity.

In certain contexts, however, it is helpful to consider a concentration as a quantity—i.e. when dealing with a fixed volume of sample, as is so often done in analyzing and interpreting electrolyte changes in body fluids. In acid-base physiology it is particularly helpful to emphasize the difference between such a group of entities as titratable acid (TA), $[HCO_3^-]$ and $[O_2Hb]$ and the intensity factors they are associated with—pH, P_{CO_2} and P_{O_2} respectively.

In other contexts, particularly in the field of fluid and electrolyte balance, a concentration can be considered an intensity factor as distinct from a total quantity of solute. The clinical importance of distinguishing between the (small) concentration of potassium in the plasma and the (large) quantity of K^+ in the body cells is known to every physician who treats electrolyte disorders. Similarly the amount of sodium in the body is becoming appreciated as a concept totally different from its plasma concentration.

Failure to appreciate these distinctions frequently leads to under- or overtreatment of patients with electrolyte disturbances. One of the commonest errors in the therapy of potassium-deficient states is to consider a near normal serum potassium level an index of a normal quantity of K^+ in the cells and to fail to administer enough potassium to repair the deficiency. A common cause of "fluid overload" in post-surgical patients is probably the failure to distinguish between dilutional hyponatremia and the loss of total body sodium.

In the acid-base field such confusion is aggravated by terminological problems. In the half century since acid-base chemistry was formulated, the word "acidosis" as used in clinical medicine has ambiguously signified either

---

* Kildeberg has recently pointed out the relation between the quantity of titratable acid and the acid intensity as follows:

"Expressed in absolute milliequivalents (= millimoles of titratable hydrogen ion), the TA is energetically an extensive quantity. Taking the pH to represent the corresponding intensive acidity measure, the product of TA (in meq × F = millicoulombs) and pH (in millivolts) may be conceived as an energy quantity."[312]

an intensity factor (measured by pH) or a quantity factor (measured by [HCO_3^-] or an approximation thereof). To physicians concerned with respiration, the distinction between intensity and quantity is fairly obvious: the pressure of a gas is qualitatively different from a quantity of it in a given volume of tissue. In other branches of medicine it is not so obvious. It is interesting that a respiratory physiologist was chosen to speak for a group of scientists wishing to clarify acid-base terminology; Campbell wrote:

> "Many people regard anything to do with the meaning of words as a waste of time and refer to 'semantics' with strong pejorative overtones. We apologize to those who have this feeling but would like to point out that in some fields progress is implemented when communication is improved . . . These are the recommendations which we would make. . . . 'Acidosis' and 'alkalosis' describe abnormal processes or conditions which would cause a deviation of pH if there were no secondary changes in response to the primary etiologic factor. . . . This practice avoids the use of the terms 'acidosis' and 'alkalosis' to describe pH deviations of the blood *per se*."[309]

The ambiguous use of these terms has so blurred the distinction between a quantity of acid and a pH change that 50 years after acid-base chemistry was founded, letters[313] to a leading clinical journal still reveal the practical consequences of losing sight of this distinction—namely, being worried about the *quantity* of acid being infused in an intravenous glucose solution stabilized at pH 5.0. (Among the reasons for retaining pH as the measure of acidity rather than substituting [H^+], whose normal value in blood plasma is 40 nM/L, is that pH at least does not "look like" a quantity or a concentration.)

To indicate that these considerations are receiving attention in clinical medicine, and to emphasize the importance of distinguishing an intensity (or potential) and a quantity (or concentration), the concluding sentences from a recent editorial in another leading clinical journal are quoted: "In general, then, the tendency for hydrogen ion to participate in chemical reactions—and assuredly physiological reactions as well—is directly determined by the chemical potential of the hydrogen ion. The pH of a system is truly an object of physiological interest in contrast to a less useful hydrogen ion 'concentration,' which itself is derived by means of a less accurate calculation from the chemical potential."[75]

It is perhaps helpful to note how three physiologic measurements of intensity (discussed in Chapters 3, 1 and 4, respectively) are related to ratios between quantities:

$$pH \propto \frac{\text{Concentration of base}}{\text{Concentration of acid}}$$

$$P_{CO_2} \propto \frac{\text{Volume of } CO_2 \text{ output}}{\text{Volume of effective ventilation}}$$

$$P_{\bar{v}O_2} \propto \frac{\text{Supply of oxygen}}{\text{Demand for oxygen}}$$

Appendix III

....

THE BASE EXCESS CONCEPT IN HISTORICAL PERSPECTIVE

THE Base Excess concept is of interest to four groups of persons. In the first are those who wish to use the BE to calculate how much bicarbonate to give to a patient with "acidosis" and who are not concerned with what BE measures nor with how it was derived. In the second are those who are satisfied by the definition: BE is the number of milliequivalents of acid or base needed to titrate one liter of blood to pH $= 7.40$ at $37°C$ while the P_{CO_2} is held constant at 40 mm Hg—with minor corrections taking account of the difference between the buffering capacity of reduced and oxygenated hemoglobin. Those in the third group understand the concept, how it arose, how it measures not just the bicarbonate base but all other buffer bases in the blood and how relatively unimportant these other bases are, compared with bicarbonate in acid-base diagnosis and treatment.* The fourth group consists of a large number of students and physicians who are not sure of the meaning of BE or of a deviation of serum bicarbonate measurement above or below a normal value of 24 mEq/L, $(\Delta[HCO_3^-])$, but who have been attracted to the BE concept because it clarifies, in some ways, the meaning of a $\Delta[HCO_3^-]$. This appendix is addressed to the fourth group.

Since the pioneer work of Van Slyke,[307] physicians throughout the world have used the serum CO_2 content (or combining power, or bicarbonate concentration) as an index of metabolic acidosis. The lower the serum CO_2 content the more severe the acidosis, roughly speaking. The Base Excess is simply a somewhat more accurate index of the severity of non-respiratory (metabolic) acidosis. Under *in vitro* (and under most *in vivo*) conditions, it precisely measures the quantity of non-carbonic (fixed) acid added to or removed from blood. It is thus an index of the "non-respiratory" component of an acid-base disturbance.

Bicarbonate concentration, $[HCO_3^-]$, can be changed by P_{CO_2} changes alone; it also can be changed by P_{CO_2} when this occurs secondarily to a non-respira-

* In contrast to their great importance in CO_2 and O_2 transport.

tory disturbance. A non-respiratory index is meant to correct for the effect on [HCO_3^-] of such a secondary P_{CO_2} change. The Base Excess index does even more than this for it takes account of the contribution of non-bicarbonate buffers. As we will see, to accomplish this involves titrating blood (graphically) to a $P_{CO_2} = 40$ and $pH = 7.4$.

Since the first attempt[40] to find such a non-respiratory index, many have tried to "correct" for the fact that the respiratory system almost always participates in acid-base regulation. Two such corrections were types of corrected pH, i.e. a pH which would have been found if P_{CO_2} were fixed: Hasselbalch's original "reduced hydrogen ion concentration,"[40] and a similar index called the eucapneic pH.[314] Neither of these indices has caught on and most efforts have been to devise an index based on the concept of a corrected bicarbonate.

The base excess is such an index. It is perhaps best understood by relating it to three other indices—the CO_2 combining power of Van Slyke, the standard bicarbonate of Jorgensen and Astrup, and the buffer base of Singer and Hastings. (See Table A2.) All four indices are similar because they reflect

TABLE A2. *"Metabolic" indices*

Index	Normal Values mEq/L	Author
Plasma bicarbonate	24 ± 2	Henderson, 1908[25]
CO_2 combining power	24 ± 3	Van Slyke, 1917[307]
Plasma CO_2 content	25 ± 2	Van Slyke, 1921[301]
Standard bicarbonate	23 ± 2	Jorgensen & Astrup, 1957[315]
Base excess	0 ± 2	Astrup et al., 1960[22]
(Base excess + 24)	24 ± 2	

The considerable range of normal and the near numerical equality of the first four indices with the quantity BE + 24 are in contrast to the variety of conceptions the names of these indices suggest. The base excess is the deviation of "buffer base" from normal.

changes in acid quantity rather than acid intensity (which is given by the pH), involve manipulating blood outside the body (i.e. titrating with CO_2) and using nomograms suitable for interpreting acid-base changes *in vitro*.* They are therefore derived measurements. Such derived measurements, although

---

* The errors which may result from the use of *in vitro* nomograms in interpreting *in vivo* changes[43] need not obscure this attempt to clarify the base excess concept. Indeed the significance of these errors (which are caused by HCO_3^- crossing blood vessel walls) cannot be appreciated unless the chemical meaning of base excess is first understood (which includes understanding how HCO_3^- crosses the red cell membrane).

often convenient and sometimes essential in physiology, require more under-standing for their proper interpretation than more direct measurements.[44]

Why use such derived acid-base measurements? Because measurements like pH and P_{CO_2} give no direct indication of the quantity of fixed acid gain or loss.

The chemical quantity in blood which changes most* with fixed acid gain or loss is the plasma bicarbonate, symbolized as $[HCO_3^-]$. This plasma bicar-bonate always decreases when fixed acid increases. But it may also decrease when P_{CO_2} decreases. Hence $[HCO_3^-]$ may not exactly reflect fixed acid quan-tity. Certain corrections, however, can be applied to the actual $[HCO_3^-]$ by manipulating blood *in vitro* and indirectly arriving at a derived value which more correctly measures fixed acid or base excess.

Before explaining how fixed acid changes can be accurately assessed *in vitro* by indirect means, it is helpful to have a clear idea of blood bicarbonate levels *in vivo*. As seen in Figure A1, $[HCO_3^-]$ is higher in plasma than in the red blood cells; in whole blood, $[HCO_3^-]$ has an intermediate value. Also seen

---

* $[H^+]$ may change, percentage-wise, even more than $[HCO_3^-]$; however, $[H^+]$ is not a quantity[5] in the same sense as a quantity of acid or base. It is only "significant as an *index* of the state of an equilibrium in which the hydrogen ion itself has little actual physical significance."[31]

The H⁺-binders in one liter of blood

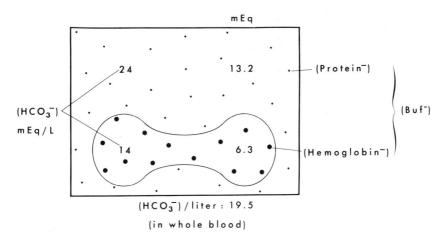

FIG. A1. The normal distribution of H⁺-binders or buffer bases (some of which are here denoted Buf⁻) in arterial blood *in vivo* when pH = 7.4 and P_{CO_2} = 40 mm Hg. If the hematocrit is 45%, the quantity of HCO_3^- in cells is 0.45 × 14 = 6.3 mEq per liter of blood. Blood *in vitro*, if collected anaerobically and without temperature change is identical to blood *in vivo* until metabolically active components (mainly the white cells) raise P_{CO_2} and lower pH and P_{O_2} or until CO_2 is allowed to enter or leave the blood.

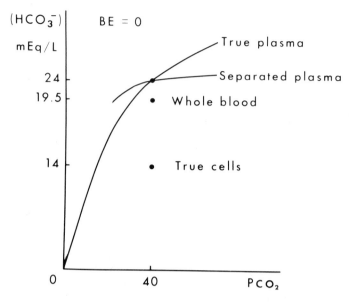

FIG. A2. True plasma is plasma in contact with red cells. When no fixed acid or base has been added, there is no excess of base in any blood component; therefore BE = 0. When only CO_2 is added (increasing Pco_2 above 40), HCO_3^- is generated (more in true than in separated plasma) but BE remains zero because no fixed acid was added. (At very low Pco_2, bicarbonate is converted to carbonate in separated plasma; hence the curve is left incomplete.)

is the fact that other H^+-binders are present. These are considered together as a concentration in whole blood labeled $[Buf^-]$ and include all non-bicarbonate buffers (such as proteins, hemoglobin and phosphates).

In Figure A2, the bicarbonate levels in different blood compartments at $Pco_2 = 40$ mm Hg are shown (closed circles) as well as the way plasma levels change with changes in Pco_2. Of the two CO_2-titration curves, the steeper is that of true plasma. This steepness results from the fact that, when CO_2 is added to blood, more HCO_3^- is generated inside than outside cells; HCO_3^- migrates from cells to true plasma, thus increasing the plasma concentration rapidly. Separated plasma (no cells present) is much less well buffered and bicarbonate generation and destruction by Pco_2 changes are much less.

With this introduction, we now will describe how artificially induced Pco_2 changes (i.e. CO_2 titrations) can be used to derive indices of fixed acid excess or loss.

The CO_2-Combining Power

In 1917, Van Slyke and Cullen[307] introduced the "CO_2 capacity," often called the CO_2-combining power. For half a century it was the most widely used index of an acid-base disturbance because its measurement did not require an arterial puncture or special precautions to prevent CO_2 from escaping from the blood. A venous blood sample was placed as for most "blood chemistries" into an open tube, taken to the laboratory and equilibrated at room temperature (after removing the red cells) with 5.5% CO_2 in air so that the Pco_2 was brought to about 40 mm Hg. Thereafter the total CO_2 content (of the separated plasma) was determined and the result reported as CO_2-combining power. This procedure had the practical advantage that venous serum or plasma could be approximately "arterialized" by bringing it to a standard Pco_2 of 40 mm Hg. See Figure A3.

Four years later, Van Slyke,[310] in discussing "means for determining state of the acid-base balance," omitted reference both to his CO_2-combining power and Hasselbalch's index, simply stating that "it is necessary to ascertain two of the involved variables" (pH, $[HCO_3^-]$, pCO_2). He specifically wrote that a similar index of available alkali (the CO_2 content of the whole blood determined after equilibration with air containing CO_2 at 40 mm tension) "is

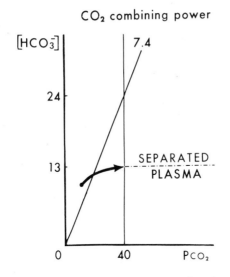

CO₂ combining power

FIG. A3. The determination of CO_2 combining power requires the exposing of separated plasma to CO_2 to bring its Pco_2 to 40 mm Hg at room temperature. The process is shown for a case of "metabolic" acidosis in which the blood in an open tube has lost CO_2 so that the plasma separated from this blood has a low Pco_2 in its initial state (solid circle) and a high pH. Adding CO_2 increases Pco_2 and the state moves to the right (crossing the pH = 7.4 line) until $Pco_2 = 40$ is reached. At this state the [Total CO_2] is determined in the separated plasma and reported as CO_2-combining power.

In the example given the [Total CO_2] was 14.8 mM/L, giving a bicarbonate of $14.8 - 0.046 \times 40 = 13$, where 0.046 is the solubility coefficient of CO_2 at room temperature in mM/L per mm Hg.[4]

inadequate . . . . as the single blood determination" and showed how such a standardized bicarbonate can be misleading. Finally, the originator of the CO_2-combining power summed up its 15 year history by stating that it "has been replaced by the simpler procedure of determining the CO_2 content of true plasma or serum which has been drawn and centrifuged with precautions to prevent loss of CO_2."[28] This procedure yields the "serum CO_2" currently reported by most clinical laboratories and is by far the most widely used index of acid gain or loss today. It is about twenty nineteenths* of the serum or plasma $[HCO_3^-]$ and is a good approximation of the "metabolic component." See Table A1.

Nevertheless the search for a corrected or non-respiratory bicarbonate has continued.

The Standard Bicarbonate

In 1957 Jorgensen and Astrup,[315] by simply equilibrating oxygenated whole blood at $37°C$ in a tonometer with gas whose $Pco_2 = 40$, reading the resulting pH, and calculating the resulting plasma bicarbonate, named this derived measurement the "standard bicarbonate." (See Fig. A4.) It is true that this procedure of bringing the Pco_2 of a blood sample to 40 mm Hg standardizes or corrects the actual bicarbonate to a new value under the special *in vitro* condition in which only the gases O_2 and CO_2 enter or leave the original blood sample. But the resulting standard bicarbonate does not represent what the plasma bicarbonate would have been if blood had remained in the body and had been brought to a Pco_2 of 40, because blood vessels are permeable to HCO_3^- as well as to free or dissolved CO_2.[72] Even *in vitro* the standard bicarbonate does not measure the magnitude of non-respiratory disturbances because it neglects the effect of buffering by HCO_3^- inside red cells. In the words of one of its originators, it therefore suffers from "the drawback that it does not show directly the amount, in mEq per liter blood, or fixed acid or base causing a change in the base content of a blood sample. . . . Here the use of change in buffer base or of base excess is helpful."[22]

The Buffer Base

This is simply the sum, in one liter of whole blood, of all bases capable of binding H^+ in the range of body pH's. For convenience these bases are considered, as in previous treatments,[23,32] to be of two kinds only, bicarbonate (in whole blood) and non-bicarbonate and written as

$$BB = [HCO_3^-] + [Buf^-]$$

* $[$ Total $CO_2] = [HCO_3^-] + [$ Dissolved or Free $CO_2]$. In normal arterial blood plasma these values are 25.2 = 24 + 0.03 × 40, where 0.03 is the solubility coefficient of CO_2 at body temperature; $25.2/24 \cong 20/19$.

Standard Bicarbonate

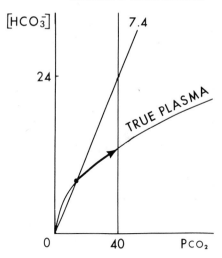

FIG. A4. For the determination of standard bicarbonate, whole blood rather than separated plasma is brought to a Pco₂ of 40 mm Hg. The pH is read and the [HCO₃⁻] of true plasma is calculated and reported as standard bicarbonate. In the example illustrated, a large increase from "actual" (solid circle) to standard bicarbonate is shown for clarity—larger than usual because the initial state of metabolic acidosis was completely compensated, i.e. the Pco₂ had been so lowered by the respiratory system that the pH had been restored to 7.4. The standard bicarbonate was devised to correct for this confusing* "respiratory disturbance" by raising the bicarbonate to a value which would correctly reflect the amount of fixed acid in the blood. As pointed out by Astrup,[22] it does not do this exactly.

where [Buf⁻] represents all non-bicarbonate bases inside and outside red cells (such as $HPO_4^=$, Protein⁻, Hb⁻, HbO_2^-, etc.).

What is required is a measure of the change in BB from normal (i.e. a change produced by a fixed acid disturbance which changed the pH and Pco₂ from 7.4 and 40 mm Hg). This change can be written

$$\Delta BB \text{ (at blood pH \& Pco}_2) = \Delta[HCO_3^-] \text{ (actual)} + \Delta[Buf^-]$$

where $\Delta[HCO_3^-]$ (actual) is the difference between the actual plasma bicarbonate and 24 mEq/L. Since [Buf⁻] itself is not directly measurable, what is done is to correct the actual bicarbonate by titrating blood with CO₂ to pH = 7.4 (graphically). This restores [Buf⁻] to normal, making $\Delta[Buf^-] = 0$, in which case

$$\Delta BB = \Delta[HCO_3^-] \text{ (corrected)}.$$

To understand this titration, consider two kinds of acid-base disturbances, respiratory alkalosis and metabolic acidosis. When CO₂ is removed from *in*

* This situation is confusing mainly because of the words used to describe it.

vitro blood (as in a tonometer), BB does not change because "*in vitro* respiratory alkalosis" lowers $[HCO_3^-]$ but raises $[Buf^-]$ by exactly the same amount:

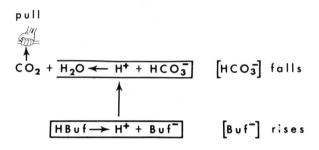

pull

$$CO_2 + \boxed{H_2O \leftarrow H^+ + HCO_3^-} \qquad [HCO_3^-] \text{ falls}$$

$$\boxed{HBuf \rightarrow H^+ + Buf^-} \qquad [Buf^-] \text{ rises}$$

Now suppose that fixed acid (say lactic acid) has been added to blood in the body and we are required to deduce how much was added, given a sample of the blood. The BB must have fallen because metabolic acidosis lowered both $[HCO_3^-]$ and $[Buf^-]$:

$$CO_2 + \boxed{H_2O \leftarrow H^+ + HCO_3^-} \qquad [HCO_3^-] \text{ falls}$$

$$\text{push} \rightarrow HLac \rightarrow H^+ + Lactate^-$$

$$\boxed{HBuf \leftarrow H^+ + Buf^-} \qquad [Buf^-] \text{ falls}$$

The amount of lactic acid which must have been added can be deduced by measuring ΔBB. To measure ΔBB in the blood sample, the sample is subjected to a controlled *in vitro* respiratory alkalosis (i.e. titrated with CO_2 by removing CO_2). If this CO_2 titration is carried out until pH = 7.4 (artificial compensation),* the $\Delta[HCO_3^-]$ (corrected) measures the lactic acid added.

The Definition and Determination of Base Excess

The base excess (BE) is defined as the deviation of BB from normal (ΔBB). It is equal to the number of milliequivalents of fixed acid or base needed to restore the pH of 1.0 liter of whole blood to 7.4 at 37°C while the P_{CO_2} is held constant at 40 mm Hg.

The BE can be measured by actually titrating with fixed acid or base, but

---

* For many clinicians the BE is best defined as the change from normal which would have been found in the whole blood bicarbonate if the metabolic acidosis or alkalosis had been completely compensated by changes in pulmonary ventilation. For chemists it is clear that when pH is brought to 7.4 by CO_2 titration, the ratio $[Buf^-]/[H\ Buf]$ is brought to a standard state which could be called normal. Hence $[Buf^-]$ is in this state; hence the $[HCO_3^-]$ resulting from the CO_2 titration quantitatively reflects fixed acid or base excess.

is much more easily either calculated (by Equation 10 on page 48 of Siggaard-Andersen's book[62]) or determined by use of the Siggaard-Andersen curve nomogram[316] or alignment nomogram.[20]

The Understanding of the Base Excess

Like any quantitative measurement, the BE can be most thoroughly understood by understanding the operations (chemical, mathematical and nomographic) needed to arrive at it. Unfortunately, none of these are simple operations for the reason that blood is a complex heterogeneous system. (For example, when fixed acid is added to blood, some of it is buffered by intra-erythrocytic HCO_3^- which has to be taken account of; when CO_2 is removed from blood in a CO_2 titration, more HCO_3^- is destroyed inside than outside red cells, and bicarbonate tends to move into the red cells from the plasma.) It is, however, possible to gain insight into the chemistry of the BE by following a CO_2 titration which corrects a measured $[HCO_3^-]$ in separated plasma as in Figure A5 and then noting that a similar CO_2 titration in whole blood

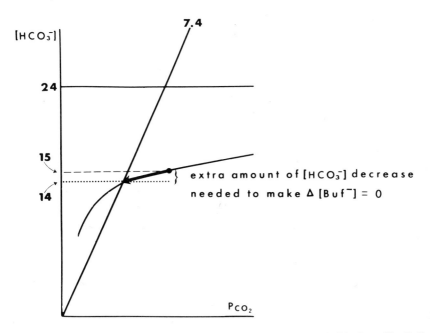

FIG. A5. In this example of metabolic acidosis (solid circle), the actual $[HCO_3^-]$ was 15 mEq/L. But this does not exactly reflect the fixed acid excess because some of the acid had been buffered by $[Buf^-]$. How much? This is easily determined in separated plasma (illustrated by the curve in the figure) by lowering the P_{CO_2} until pH = 7.4, because at this pH $\Delta [Buf^-] = 0$ and the ΔBB (or BE) is measured by the difference between *14* and 24 (i.e., −10) rather than 15 and 24 (i.e. −9).

involves not one but two corrections* of the actual $[HCO_3^-]$ of true plasma as in Figure A6. In both these examples the graphic removal of CO_2 is shown to lower (i.e. correct) the actual $[HCO_3^-]$ by just the amount needed to make $\Delta[Buf^-] = 0$ and they illustrate the same processes discussed above under buffer base. Figure A7 is another way of illustrating how the CO_2 titration can be carried out graphically, but the fact that the end result is to change the actual bicarbonate to a corrected value is not so evident.

The Clinical Significance of the Base Excess

Figure A8 shows a typical example of a respiratory disturbance. It can be seen that there is very little difference between BE and $[HCO_3^-] - 24$ mEq/L except when the Pco_2 was very high. In other words, the actual $\Delta[HCO_3^-]$ and the corrected $\Delta[HCO_3^-]$ ($=BE$) are very little different.

---

* The need for two corrections of the actual $[HCO_3^-]$ is most clearly seen by dissecting Equation 16 (page 49, in Siggaard-Andersen[62]).

Base Excess

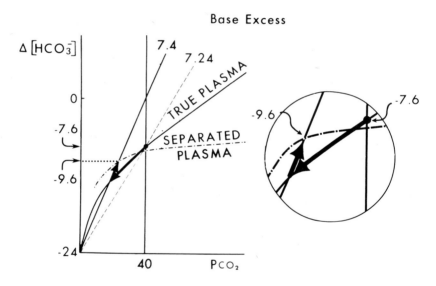

FIG. A6. The BE concept explained on the classical $Pco_2 - [HCO_3^-]$ diagram with the ordinate changed to $\Delta[HCO_3^-]$. An actual case of metabolic acidosis of $[HCO_3^-] = 16.4$, $Pco_2 = 40$, pH = 7.24 is plotted as a solid circle. To determine BE (to correct $\Delta[HCO_3^-]$ from -7.6 to -9.6), first lower Pco_2 until the circle moves to the intersection of the true plasma curve with the iso-pH line of 7.4 (long first correction arrow). Then raise the Pco_2 and simultaneously replace the red cells with plasma (graphically) by moving up the pH = 7.40 line until it intersects with the separated plasma curve. Read off the corrected $\Delta[HCO_3^-]$ as -9.6 on the ordinate. This is the base excess. The graphic replacement of red cells by plasma is further explained in the legend to Figure A7.

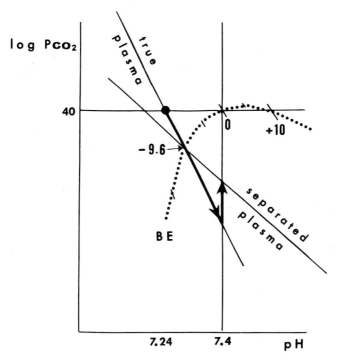

FIG. A7. As in Figure A6, the initial acid-base disturbance is indicated by a solid circle at $[HCO_3^-] = 16.4$, $Pco_2 = 40$, pH $= 7.24$. This diagram has no axis specifically showing the actual bicarbonate. The process of graphic titration is the same as before: lower Pco_2 in whole blood until pH $= 7.4$, and then raise it to keep pH constant as red cells are replaced by plasma (i.e. moving between separated plasma and true plasma lines).

The curved dotted line is "the geometric locus of the acid-base values for which $BE_c = BE_p$. BE_c is defined as the titratable base on titration of the erythrocyte fluid at $Pco_2 = 40$ mm Hg, to the pH value found in the erythrocytes when the plasma pH is 7.40."[62]

Graphically replacing red cells with plasma at the same pH has the effect of raising the plasma bicarbonate level to a point reflecting the whole blood bicarbonate change; it therefore takes account of the contribution of intra red cell bicarbonate.

The small magnitude of this difference—seldom more than 3 mEq/L even in very ill patients—is of small consequence in planning therapy to correct an acidosis with bicarbonate solution or an alkalosis with ammonium chloride.

For practical purposes Δ actual $[HCO_3^-]$ can be used just as well as BE in the formula (as it is usually written):

$$\text{mEq acid or base needed} = 0.3 \times BE \times \text{body wt.}$$

Similarly, following the plasma bicarbonate level as it rises above and falls below 24 mEq/L is all that is needed to "assess the metabolic component."

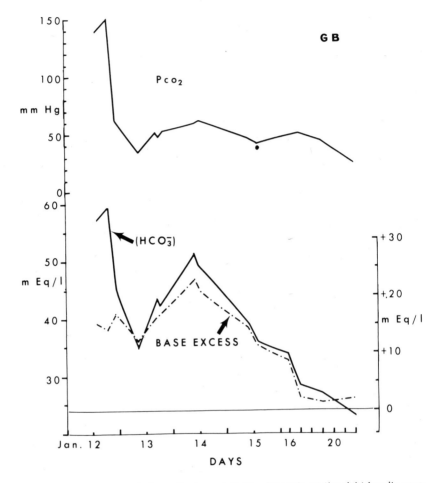

FIG. A8. The bicarbonate scale is on the lower left. The BE scale on the right has its zero point opposite [HCO$_3^-$] = 24 mEq/L. GB had been on continuous O$_2$ therapy at home before admission to the hospital. His "normal" Pco$_2$ was about 60 and his normal [HCO$_3^-$] 40 mEq/L. The acute exacerbation of his bronchitis raised his Pco$_2$ to 100 mm Hg and his [HCO$_3^-$] by 20 mEq/L—as expected from his CO$_2$ dissociation curve. Only with very large Pco$_2$ change does the [HCO$_3^-$] deviate much from BE + 24 mEq/L.

This statement may disappoint the reader who has had the patience to follow the present detailed attempt to explain derived non-respiratory indices, for it merely echoes the conclusion that "traditional measurements of pH, Pco$_2$ and plasma bicarbonate concentration continue to be the most reliable biochemical guides in the analysis of acid-base disturbances."[61] The foregoing explanations are offered, nevertheless, to those who may wish to understand the Base Excess as a chemical measurement before accepting or rejecting it for use in medicine.*

Recommendations

It is to be noted that the general concept of BE as a quantity factor need not involve the pH to any great extent. Because blood pH is the fundamental measurement on which all other acid-base parameters depend and which profoundly affects most physiologic functions, the chief danger in focusing too intently on quantity factors such as BE is that the critical importance of pH and other intensity factors is forgotten. The deceptively simple scale (plus—0—minus) has become, too often, a signal for action, and numbers on this scale, instead of guides to intelligent therapy, have become part of a rule of thumb. In other words, clinicians should be asking if a pH change is worth correcting rather than feeling obligated to correct a BE greater or less than zero.

For these reasons it is recommended that laboratory slips containing acid-base information be drawn up as follows:

$$pH\text{———}\qquad [HCO_3^-]\text{———}$$
$$P_{CO_2}\text{———}\qquad BE\text{———}$$
$$P_{O_2}\text{———}\qquad \% \text{ Sat}\text{———}$$

This form will allow the quantitative relationship of $[HCO_3^-]$ to BE to be revealed. It also has the advantage of separating intensity factors (on the left) and quantity factors (on the right).

---

*When such understanding is achieved, errors arising from the application of *in vitro* derivations to *in vivo* circumstances can be avoided. These errors are particularly large in acute respiratory failure with hypercapnia because HCO_3^- moves from plasma to extracellular fluid causing a falsely "negative base excess."[43] (See pp. 34–38.)

References

1. HALDANE, J. S.: Respiration. New Haven, Yale University Press, 1922, pp. 67–68.
2. HENDERSON, L. J.: The Fitness of the Environment. New York, Macmillan Co., 1913, pp. 139, 140.
3. HENDERSON, L. J.: Blood. A Study in General Physiology. New Haven, Yale University Press, 1928, pp. 32, 55, 56, 78, 82.
4. SEVERINGHAUS, J. W.: Blood gas concentrations. *In* Handbook of Physiology, Washington, D.C., Amer. Physiol. Soc. Sec. 3, Vol. II, 1965, Ch. 61, p. 1475.
5. BELL, R. P.: The Proton in Chemistry. Ithaca, N.Y., Cornell University Press, 1959, pp. 3, 9, 10, 30.
6. EDSALL, J. T., and J. WYMAN: Biophysical Chemistry, Vol. 1. New York, Academic Press, Inc., 1958, pp. 411, 452.
7. SCHWARTZ, W. B., N. C. BRACKETT, JR., and J. J. COHEN: The response of extracellular hydrogen ion concentration to graded degrees of chronic hypercapnia: the physiologic limits of the defense of pH. J. Clin. Invest. *44:* 291, 1965.
8. ROBIN, E. D., P. A. BROMBERG, and F. S. TUSHAN: Carbon dioxide in body fluids. New Eng. J. Med. *280:* 162, 1969.
9. SAUNIER, C., M. C. AUG-LAXENAIRE, M. SCHIBI, and P. SADOUL: Acid-base and electrolyte equilibrium of arterial blood and cerebrospinal fluid in respiratory insufficiency. Respiration *26:* 81, 1969.
10. POSNER, J. B., A. G. SWANSON, and F. PLUM: Acid-base balance in cerebrospinal fluid. Arch. Neurol. *12:* 479, 1965.
11. ALROY, G. G., and D. C. FLENLEY: The acidity of the cerebrospinal fluid in man with particular reference to chronic ventilatory failure. Clin. Sci. *33:* 335, 1967.
12. MITCHELL, R. A., C. T. CARMAN, J. W. SEVERINGHAUS, B. W. RICHARDSON, M. M. SINGER, and S. SHNIDER: Stability of cerebrospinal fluid pH in chronic acid-base disturbances in blood. J. Appl. Physiol. *20:* 443, 1965.
13. CAIN, C. C., and A. B. OTIS: Some physiological effects resulting from added resistance to respiration. J. Aviation Med. *20:* 149, 1949.
14. RILEY, R. L. (Editorial): The work of breathing and its relation to respiratory acidosis. Ann. Intern. Med. *41:* 172, 1954.
15. CHERNIACK, R. M.: Work of breathing and the ventilatory response to CO_2. *In* Handbook of Physiology, Washington, D.C., Amer. Physiol. Soc., Sec. 3, Vol. II, 1965, Ch. 60, p. 1469.

16. PHILLIPS, B., and D. I. PERETZ: A comparison of central venous and arterial blood gas values in the critically ill. Ann. Intern. Med. *70:* 745, 1969.

17. SHAW, J. C.: Arterial sampling from the radial artery in premature and full-term infants. Lancet *2:* 389, 1968.

18. COTTON, E. K.: Personal communication, 1969.

19. VAN SLYKE, D. D., and J. SENDROY: Studies of gas and electrolyte equilibria in blood. XV. Line charts for graphic calculations by Henderson-Hasselbalch equation, and for calculating plasma carbon dioxide content from whole blood content. J. Biol. Chem. *79:* 781, 1928.

20. SIGGAARD-ANDERSEN, O.: Blood acid-base alignment nomogram. Scales for pH, P_{CO_2}, base excess of whole blood of different hemoglobin concentrations, plasma bicarbonate and plasma total CO_2. Scand. J. Clin. Lab. Invest. *15:* 211, 1963.

21. SINCLAIR, M. J., R. A. HART, H. M. POPE, and E. J. M. CAMPBELL: The use of the Henderson-Hasselbalch equation in routine medical practice. Clin. Chim. Acta *19:* 63, 1968.

22. ASTRUP, P., K. JORGENSEN, O. SIGGAARD-ANDERSEN, and K. ENGEL: The acid-base metabolism. A new approach. Lancet *1:* 1035, 1960.

23. WINTERS, R. W., K. ENGEL, and R. B. DELL: Acid-Base Physiology in Medicine. Cleveland, The London Co., 1967, pp. 44–48, 279–286.

24. COMROE, J. H., JR.: The peripheral chemoreceptors. *In* Handbook of Physiology, Washington, D.C., Amer. Physiol. Soc., Sec. 3, Vol. 1, 1964, Ch. 23, p. 557.

25. HENDERSON, L. J.: The theory of neutrality regulation in the animal organism. Amer. J. Physiol. *21:* 427, 1908.

26. DYBKAER, R.: Quantities and units in clinical chemistry. Scand. J. Clin. Lab. Invest. *25:* 5, 1970.

27. PITTS, R. F.: Physiology of the Kidney and Body Fluids, 2nd Ed. Chicago, Year Book Medical Publishers, Inc., 1968, pp. 37, 170, 180, 182, 188.

28. PETERS, J. P., and D. D. VAN SLYKE: Quantitative Clinical Chemistry, Vol. I, Interpretations. Baltimore, Williams & Wilkins Co., 1931, pp. 539, 873, 875, 894, 925, 935.

29. ROUGHTON, F. J. W.: Transport of oxygen and carbon dioxide. *In* Handbook of Physiology, Washington, D.C., Amer. Physiol. Soc., Sec. 3, Vol. I, 1964, Ch. 31, p. 767.

30. HENDERSON, L. J., A. V. BOCK, H. FIELD, JR., and J. L. STODDARD: Blood as a physicochemical system. J. Biol. Chem. *59:* 379, 1924.

31. CLARK, W. M.: The Determination of Hydrogen Ions, 3rd Ed. Baltimore, Williams & Wilkins Co., 1928, pp. xi, xiii, 35–39, 40, 41, 520.

32. SINGER, R. B., and A. B. HASTINGS: An improved clinical method for the estimation of disturbances of the acid-base balance of human blood. Medicine *27:* 223, 1948.

33. MICHEL, C. C.: The buffering behavior of blood during hypoxaemia and respiratory exchange: theory. Resp. Physiol. *4:* 283, 1968.

34. REHM, W. S.: Ion permeability and electrical resistance of the frog's gastric mucosa. Fed. Proc. *26:* 1303, 1967.

35. BRODSKY, W. A., and T. P. SCHILB: Mechanism of acidification in turtle bladder. Fed. Proc. *26:* 1314, 1967.

36. WOODBURY, D. M.: Physiology of Body Fluids. *In* Physiology and Biophysics (Ed. by T. C. Ruch and H. D. Patton). Philadelphia, W. B. Saunders & Co., 1965.

37. BURNELL, J. M.: *In vivo* response of muscle to changes in CO_2 tension or extracellular bicarbonate. Amer. J. Physiol. *215:* 1376, 1968.

38. SWAN, R. C., and R. F. PITTS: Neutralization of infused acid by nephrectomized dogs. J. Clin. Invest. *34:* 205, 1955.

39. SCHWARTZ, W. B., K. J. ORNING, and R. PORTER: The internal distribution of hydrogen ions with varying degrees of metabolic acidosis. J. Clin. Invest. *36:* 373, 1957.

40. HASSELBALCH, K. A.: Die "reduzierte" und die "regulierte" Wasserstoffzahl des blutes. Biochem. Z. *74:* 56, 1916.

41. CHRISTIANSEN, J., C. G. DOUGLAS, and J. S. HALDANE: The adsorption and dissociation of carbon dioxide by human blood. J. Physiol. *48:* 244, 1914.

42. MORRIS, M. E., and R. A. MILLAR: Blood pH/plasma catecholamine relationships: respiratory acidosis. Brit. J. Anaesth. *34:* 672, 1962.

43. STERN, L. I., and D. H. SIMMONS: Estimation of non-respiratory acid-base disturbances. J. Appl. Physiol. *27:* 21, 1969.

44. FILLEY, G. F.: Acid-base language versus acid-base measurements. Pediatrics *43:* 830, 1969.

45. COHEN, J. J., N. C. BRACKETT, JR., and W. B. SCHWARTZ: The nature of the carbon dioxide titration curve in the normal dog. J. Clin. Invest. *43:* 777, 1964.

46. LASZLO, G., T. J. H. CLARK, and E. J. M. CAMPBELL: The immediate buffering of retained carbon dioxide in man. Clin. Sci. *37:* 299, 1969.

47. KURTZMAN, N. A.: Regulation of renal bicarbonate reabsorption by extracellular fluid volume. J. Clin. Invest. *49:* 586, 1970.

48. ROBINSON, J. R.: Fundamentals of Acid-Base Regulation, 3rd Ed. Springfield, Ill., Charles C Thomas, 1967.

49. SULLIVAN, W. J., and P. J. DORMAN: Renal response to chronic respiratory acidosis. J. Clin. Invest. *34:* 268, 1955.

50. BRAZEAU, P., and A. GILMAN: Effect of plasma CO_2 tension on renal tubular reabsorption of bicarbonate. Amer. J. Physiol. *175:* 33, 1952.

51. BERLINER, R. W., T. J. KENNEDY, JR., and J. ORLOFF: Relationship between acidification of urine and potassium metabolism; effect of carbonic anhydrase inhibition on potassium excretion. Amer. J. Med. *11:* 274, 1951.

52. SCHWARTZ, W. B., C. VAN YPERSELE DE STRIHOU, and J. P. KASSIRER: Role of anions in metabolic alkalosis and potassium deficiency. New Eng. J. Med. *279:* 630, 1968.

53. KASSIRER, J. P., and W. B. SCHWARTZ: Correction of metabolic alkalosis in man without repair of potassium deficiency: A reevaluation of the role of potassium. Amer. J. Med. *40:* 19, 1966.

54. PITTS, R. F., and W. D. LOTSPEICH: Bicarbonate and the renal regulation of acid-base balance. Amer. J. Physiol. *147:* 138, 1946.

55. HILTON, J. G., N. E. CAPECI, G. T. KISS, O. R. KRUESI, V. V. GLAVIANO, and R. WEGRIA: The effect of acute elevation of the plasma chloride concentration on the renal excretion of bicarbonate during acute respiratory acidosis. J. Clin. Invest. *35:* 481, 1956.

56. VAN SLYKE, D. D., and J. M. NEILL: The determination of gases in blood and other solutions by vacuum extraction and manometric measurement I. J. Biol. Chem. *61:* 523, 1924.
57. NATELSON, S.: Routine use of ultramicro methods in the clinical laboratory. Amer. J. Clin. Path. *21:* 1153, 1951.
58. SEVERINGHAUS, J. W.: Blood gas calculator. J. Appl. Physiol. *21:* 1108, 1966.
59. ASTRUP, P.: Acid-base disorders. New Eng. J. Med. *269:* 817, 1963.
60. SIGGAARD-ANDERSEN, O.: Titratable acid or base of body fluids. Ann. N.Y. Acad. Sci. *133:* 41, 1966.
61. SCHWARTZ, W. B., and A. S. RELMAN: A critique of the parameters used in the evaluation of acid-base disorders. New Eng. J. Med. *268:* 1382, 1963.
62. SIGGAARD-ANDERSEN, O.: The Acid-Base Status of the Blood, 2nd Ed. Copenhagen, Munksgaard, 1964.
63. BROWN, E. B., JR.: Acid-base debate. Anesthesiology *27:* 333, 1966.
64. MELLEMGAARD, K., and P. ASTRUP: The quantitative determination of surplus amounts of acid or base in the human body. Scand. J. Clin. Lab. Invest. *12:* 187, 1960.
65. WOODBURY, J. W.: Regulation of pH. *In* Physiology and Biophysics (Ed. by T. C. Ruch and H. D. Patton). Philadelphia, W. B. Saunders Co., 1965.
66. BADER, M. E., and R. A. BADER: Respiratory stimulants in obstructive lung disease. Amer. J. Med. *38:* 165, 1965.
67. MONCRIEF, J. A.: The status of topical antibacterial therapy in the treatment of burns. Surgery *63:* 862, 1968.
68. STEWART, J. D., and H. M. ROURKE: The effects of large intravenous infusions on body fluid. J. Clin. Invest. *21:* 197, 1942.
69. SHIRES, G. T., and J. HOLMAN: Dilution acidosis. Ann. Intern. Med. *28:* 557, 1948.
70. WINTERS, R. W., P. R. SCAGLIONE, G. G. NAHAS, and M. VEROSKY: The mechanism of acidosis produced by hyperosmotic infusions. J. Clin. Invest. *43:* 647, 1964.
71. CANNON, P. J., H. O. HEINEMANN, M. S. ALBERT, J. H. LARAGH, and R. W. WINTERS: "Contraction" alkalosis after diuresis of edematous patients with ethacrynic acid. Ann. Intern. Med. *62:* 979, 1965.
72. BROWN, E. B., JR.: Plasma electrolyte composition in dogs breathing high CO_2 mixtures; source of bicarbonate deficit in severe respiratory acidosis. J. Lab. Clin. Med. *55:* 767, 1960.
73. HILLS, A. G., and E. L. REID: pH defended—is it defensible? Ann. Intern. Med. *65:* 1150, 1966.
74. HILLS, A. G., and E. L. REID: More on pH. Ann. Intern. Med. *66:* 238, 1967a.
75. DAVIS, R. P.: Editorial: Logland: A Gibbsian view of acid-base balance. Amer. J. Med. *42:* 159, 1967.
76. FILLEY, G. F.: Acid-base regulation: classical concepts and modern measurements. Johns Hopkins Med. J. *120:* 355, 1967b.
77. WADDELL, W. J., and R. G. BATES: Intracellular pH. Physiol. Rev. *49:* 285, 1969.
78. MITCHELL, PETER: Chemiosmotic Coupling and Energy Transduction, Glynn Research Labs. Bodmin, Cornwall, England, 1968, p. 1.

79. Bronsted, J. N.: Acid and basic catalysis. Chem. Rev. *5:* 231, 1928.
80. Bent, H. A.: The Second Law. New York, Oxford University Press, 1965, pp. 47, 213.
81. Baldwin, E.: Dynamic Aspects of Biochemistry. Cambridge, Cambridge University Press, 1948.
82. Frisell, W. R.: Acid-Base Chemistry in Medicine. New York, Macmillan Co., 1968.
83. Bates, R. G.: Acids, bases and buffers. Ann. N.Y. Acad. Sci. *133:* 25, 1966.
84. Pitts, R. F., and W. J. Stone: Renal Metabolism and the Excretion of Ammonia. Proc. 3rd Int. Congr. Nephrol., Washington, 1966, Vol. I. Basel & New York, Karger, 1967, p. 123.
85. Robinson, J. R.: Renal Acid-Base Control. *In* Biochem. Clinics No. 2. Chicago, The Commerce Clearinghouse, 1963, pp. 116, 118, 119.
86. Elkinton, J. R.: Hydrogen ion turnover in health and in renal disease. Ann. Intern. Med. *57:* 660, 1962.
87. Lemann, J., Jr., and A. S. Relman: The relation of sulfur metabolism to acid-base balance and electrolyte excretion: the effects of DL-methionine in normal man. J. Clin. Invest. *38:* 2215, 1959.
88. Lennon, E. J., J. Lemann, Jr., and A. S. Relman: The effects of phosphoproteins on acid balance in normal subjects. J. Clin. Invest. *41:* 637, 1962.
89. Lemann, J., Jr., E. J. Lennon, A. D. Goodman, J. R. Litzow, and A. S. Relman: The net balance of acid in subjects given large loads of acid or alkali. J. Clin. Invest. *44:* 507, 1965.
90. Camien, M. N., L. M. Smith, T. J. Reilly, and D. H. Simmons: Determination of total cation-forming mineral elements in feces and urine and its relation to renal "net acid" excretion. Proc. Soc. Exp. Biol. Med. *123:* 686, 1966.
91. Camien, M. N., D. H. Simmons, and H. C. Gonick: A critical reappraisal of "acid-base" balance. J. Clin. Nutrition *22:* 786, 1969.
92. Steinmetz, P. R.: Acid-base relations in epithelium of turtle bladder: site of active step in acidification and role of metabolic CO_2. J. Clin. Invest. *48:* 1258, 1969.
93. Hills, A. G., and E. L. Reid: Renal maintenance of acid-base balance in health. Johns Hopkins Med. J. *120:* 368, 1967.
94. Hills, A. G.: The fitness of carbon dioxide and ammonia to serve acid-base balance: mammalian urine and plasma pH as evolutionary adaptations. Amer. Naturalist *103:* 131, 1969.
95. Elkinton, J. R., D. K. McCurdy, and V. M. Buckalew, Jr.: Hydrogen ion and the kidney. *In* Renal Disease, 2nd Ed. (D. A. K. Black, Ed.). Oxford, Blackwell Scientific Publications, Ltd., 1967.
96. Stinebaugh, B., R. B. Miller, and A. S. Relman: The influence of non-reabsorbable anions on acid excretion. Clin. Sci. *36:* 53, 1969.
97. Elkinton, J. R.: Clinical disorders of acid-base regulation. Med. Clin. N. Amer. *50:* 1325, 1966.
98. Mithoefer, J. C., M. S. Karetzky, and W. F. Porter: The *in vivo* carbon dioxide titration curve in the presence of hypoxia. Resp. Physiol. *4:* 15, 1968.

99. GILMAN, A.: The mechanism of diuretic action of the carbonic anhydrase inhibitors. Ann. N.Y. Acad. Sci. *71:* 355, 1957–1958.

100. GOLDRING, R. M., P. J. CANNON, H. O. Heinemann, and A. P. FISHMAN: Respiratory adjustment to chronic metabolic alkalosis in man. J. Clin. Invest. *47:* 188, 1968.

101. ROBINSON, J. R.: Fundamentals of Acid-Base Regulation. Springfield, Ill., Charles C Thomas, 1961.

102. CLARK, W. M.: Topics in Physical Chemistry. Baltimore, Williams & Wilkins Co., 1948, pp. 219, 241, 359.

103. BARCROFT, J.: The Respiratory Function of the Blood. Cambridge, Cambridge Univerity Press, 1914, pp. 73, 225, 227.

104. CONANT, J. B.: Harvard Case Histories in Experimental Science, Vol. I, James B. Conant, Ed. Cambridge, Harvard University Press, 1957.

105. DAVIES, P. W., and D. W. BRONK: Oxygen tension in mammalian brain. Fed. Proc. *16:* 689, 1957.

106. KETY, S. S.: Determinants of tissue oxygen tension. Fed. Proc. *61:* 666, 1957.

107. COMROE, J. H., JR.: Physiology of Respiration. Chicago, Year Book Medical Publishers, Inc., 1965, pp. 54, 55, 167–169, 210.

108. SENDROY, J., JR., R. T. DILLON, and D. D. VAN SLYKE: Studies of gas and electrolyte equilibrium in blood: Solubility and physical state of uncombined oxygen in blood. J. Biol. Chem. *105:* 597, 1934.

109. BATTINO, R., and H. L. CLEVER: The solubility of gases in liquids. Chem. Rev. *66:* 395, 1966.

110. RICHARDS, D. W., JR.: The circulation in traumatic shock in man. Harvey Lect. *39:* 217, 1943–44.

111. BARCROFT, J.: Physiological effects of insufficient oxygen supply. Nature *106:* 125, 1920–21.

112. FROESE, G.: The respiration of ascites tumour cells at low oxygen concentrations. Biochem. Biophys. Acta *57:* 509, 1962.

113. LONGMUIR, I. S.: Respiration rate of rat-liver cells at low oxygen concentrations. Biochem. J. *65:* 378, 1957.

114. BÄNDER, A., and M. KIESE: Die Wirkung des sauerstoffubertragenden Ferments in Mitochondrien aus Rattenlebern bei niedrigen Sauerstoffdrucken, Nauyn-Schmiedeberg. Arch. Exptl. Pathol. Pharmakol. *224:* 312, 1955.

115. JÖBSIS, F. F.: Basic processes in cellular respiration. *In* Handbook of Physiology, Washington, D.C., Amer. Physiol. Soc., Sec. 3, Respiration, Vol. I, 1964, Ch. 2.

116. CHANCE, B., B. SCHOENER, and F. SCHINDLER: The intracellular oxidation-reduction state. *In* Oxygen in the Animal Organism (Ed. by F. Dickens and E. Neil). New York, Macmillan Co., 1964.

117. TENNEY, S. M., and T. W. LAMB: Physiological consequences of hypoventilation and hyperventilation. *In* Handbook of Physiology, Washington, D.C., Amer. Physiol. Soc. Sec. 3, Respiration, Vol. II, 1965, Ch. 37, p. 979.

118. LÜBBERS, D. W., U. C. LUFT, G. THEWS, and E. WILZLEB (Editors): Oxygen Transport in Blood and Tissue. Stuttgart, George Thieme Verlag (New York, Intercontinental Medical Book), 1968.

119. BARBASHOVA, Z. I.: Cellular level of adaptation. *In* Handbook of Physiology,

Washington, D.C., Amer. Physiol. Soc., Sec. 4, Adaptation to the environment, 1964, Chap. 4, p. 37.

120. KREBS, H. A., and H. L. KORNBERG: Energy Transformations in Living Matter. Berlin, Springer-Verlag, 1957.

121. MASON, D. T., *et al.*: Physiologic approach to the treatment of angina pectoris. New Eng. J. Med. *281:* 1225, 1969.

122. ALLEYNE, G. A. O., *et al.*: Effect of pronethalol in angina pectoris. Brit. Med. J. *2:* 1226, 1963.

123. SRIVASTAVA, S. C., H. A. DEWAR, and D. J. NEWELL: Double-blind trial of propranolol (Inderal) in angina of effort. Brit. Med. J. *2:* 724, 1964.

124. GILLAM, P. M. S., B. N. C. PRICHARD: Use of propranolol in angina pectoris. Brit. Med. J. *2:* 337, 1965.

125. GRANT, R. H., P. KEELAN, R. J., KERNOHAN *et al.*: Multicenter trial of propranolol in angina pectoris. Amer. J. Cardiol. *18:* 361, 1966.

126. GILLAM, P. M. S., and B. N. PRICHARD: Propranolol in the therapy of angina pectoris. Amer. J. Cardiol. *18:* 366, 1966.

127. GUYTON, A. C.: Textbook of Medical Physiology, 3rd Ed. Philadelphia, W. B. Saunders Co., 1966.

128. SCHMIDT, C. F.: Cerebral blood supply and cerebral oxidative metabolism. *In* Oxygen in the Animal Organism (Ed. by F. Dickens and E. Neil). New York, Macmillan Co., 1964.

129. PLUM, F., and J. B. POSNER: Diagnosis of Stupor and Coma. Philadelphia, F. A. Davis Co., 1966, pp. 115, 119, 142.

130. RUBNER, M.: The Laws of Energy Consumption in Nutrition, Franz Deuticke, 1902. Transl. by A. Markoff and A. Sandri-White, U.S. Army Research Inst. Environmental Med., Natick, Mass., 1968.

131. NGAI, S. H., and E. M. PAPPER: Metabolic Effects of Anesthesia. Springfield, Ill., Charles C Thomas, 1962.

132. GRANDE, F.: Man under caloric deficiency. *In* Handbook of Physiology, Washington, D.C., Amer. Physiol. Soc., Sec. 4, Adaptation to environment, 1964, Ch. 59, p. 911.

133. TEPPERMAN, J.: Metabolic and Endocrine Physiology, 2nd Ed. Chicago, Year Book Medical Publishers, Inc., 1968.

134. MUSACCHIA, X. J., and J. F. SAUNDERS: Depressed Metabolism. New York, American Elsevier Pub. Co., Inc., 1969.

135. NEUFELD, A. H., C. G. TOBY, and R. L. NOBLE: Biochemical findings in normal and trauma resistant rats following trauma. Proc. Soc. Exp. Biol. Med. *54:* 249, 1943.

136. NOBLE, R. L.: The development of resistance by rats and guinea pigs to amounts of trauma usually fatal. Amer. J. Physiol. *138:* 346, 1943.

137. HRUZA, Z., and O. POUPA: Injured man. *In* Handbook of Physiology, Washington, D.C., Amer. Physiol. Soc. Sec. 4 Adaptation to Environment, 1964, Ch. 60, p. 939.

138. BARBOUR, J. H., and M. H. SEEVERS: A comparison of the acute and chronic toxicity of carbon dioxide with especial reference to its narcotic action. J. Pharmacol. Exp. Ther. *78:* 11, 1943.

139. KARETZKY, M. S., and S. M. CAIN: The effect of carbon dioxide on oxygen uptake during hyperventilation in normal man. J. Appl. Physiol. *28:* 8, 1970.

140. FILLEY, G. F.: Pulmonary ventilation and the oxygen cost of exercise in emphysema, Trans. Amer. Clin. Climat. Assoc. *70:* 193, 1958.

141. LEVISON, H., and R. M. CHERNIACK: Ventilatory cost of exercise in chronic obstructive pulmonary disease. J. Appl. Physiol. *25:* 21, 1968.

142. METCALFE, J.: A comparison of mechanisms of oxygen transport among several mammalian species. *In* Advances in Experimental Medicine and Biology, Vol. 6, Red Cell Metabolism and Function. New York, Plenum Publishing Corp., 1970.

143. REEVE, E. B., and A. C. GUYTON: Physical Bases of Circulatory Transport. Philadelphia, W. B. Saunders Co., 1967.

144. LANDIS, E. M., and J. R. PAPPENHEIMER: Exchange of substances through the capillary walls. *In* Handbook of Physiology, Washington, D.C., Amer. Physiol. Soc., Sec. 2, Circulation, Vol. II, 1963, Ch. 29, p. 961.

145. VALDIVIA, E.: Total capillary bed in striated muscle of guinea pigs native to the Peruvian mountains. Amer. J. Physiol. *194:* 585, 1958.

146. STAINSBY, W. N., and A. B. OTIS: Blood flow, blood oxygen tension, oxygen uptake and oxygen transport in skeletal muscle. Amer. J. Physiol. *206:* 858, 1964.

147. KETY, S. S., and C. F. SCHMIDT: Effects of altered arterial tensions of carbon dioxide and oxygen on cerebral blood flow and cerebral oxygen consumption of normal young men. J. Clin. Invest. *27:* 484, 1948.

148. FOLKOW, B, C. HEYMANS, and E. NEIL: Integrated aspects of cardiovascular regulation. *In* Handbook of Physiology, Washington, D.C., Amer. Physiol. Soc. Sec. 2, Circulation, Vol. III, 1965, Ch. 49, p. 1787.

149. YONCE, L. R., and B. FOLKOW: The integration of the cardiovascular response to diving. Amer. Heart J. *79:* 1, 1970.

150. KORNER, P. I.: Circulatory adaptations in hypoxia. Physiol. Rev. *39:* 687, 1959.

151. CROPP, G. J.: Cardiovascular function in children with severe anemia. Circulation *39:* 775, 1969.

152. SHAW, D. B., and T. SIMPSON: Polycythaemia in emphysema. Quart. J. Med. *30:* 135, 1961.

153. WEIL, J. V., G. JAMIESON, D. W. BROWN, and R. F. GROVER: The red cell mass-arterial oxygen relationship in normal man. J. Clin. Invest. *47:* 1627, 1968.

154. FILLEY, G. F., H. J. BECKWITT, J. T. REEVES, and R. S. MITCHELL: Chronic obstructive bronchopulmonary disease. II. Oxygen transport in two clinical types. Amer. J. Med. *44:* 26, 1968.

155. STAMATOYANNOPOULOS, G., J. T. PARER, and C. A. FINCH: Physiologic Implications of a hemoglobin with decreased oxygen affinity (Hemoglobin Seattle). New Eng. J. Med. *281:* 915, 1969.

156. CHARACHE, S., D. J. WEATHERALL, and J. B. CLEGG: Polycythemia associated with a hemoglobinopathy. J. Clin. Invest. *45:* 813, 1966.

157. RICHARDSON, T. Q., and A. C. GUYTON: Effects of polycythemia and anemia on cardiac output and other circulatory factors. Amer. J. Physiol. *197:* 1167, 1959.

158. BANCHERO, N., F. SIME, D. PEÑALOZA, J. CRUZ, R. GAMBOA, and E. MARTICORENA: Pulmonary pressure, cardiac output and arterial oxygen saturation during exercise at high altitude and at sea level. Circulation *33:* 249, 1966.

159. BARCROFT, J.: Features in the Architecture of Physiological Function. Cambridge, Cambridge University Press, 1938, p. 219.

160. FORSTER, R. E.: Diffusion of Gases. *In* Handbook of Physiology, Washington, D.C., Amer. Physiol. Soc., Sec. 3, Respiration, Vol. I, 1964, Ch. 33, p. 839.

161. CHANUTIN, A., and R. R, CURNISH: Effect of organic and inorganic phosphates on the oxygen equilibrium of human erythrocytes. Arch. Biochem. *121:* 96, 1967.

162. BENESCH, R., and R. E. BENESCH: The effect of organic phosphates from the human erythrocyte on the allosteric properties of hemoglobin. Biochem. Biophys. Res. Commun. *26:* 162, 1967.

163. LENFANT, C., J. TORRANCE, E. ENGLISH, C. A. FINCH, C. REYNAFARJE, J. RAMOS, and J. FAURA: Effect of altitude on oxygen binding by hemoglobin and on organic phosphate levels. J. Clin. Invest. *47:* 2652, 1968.

164. BREWER, G. J. (Editor): Advances in experimental medicine and biology, Vol. 6, Red Cell Metabolism and Function. New York, Plenum Publishing Corp., 1970.

165. LENFANT, C., P. WAYS, C. AUCUTT, and J. CRUZ: Effect of chronic hypoxic hypoxia on the O_2-Hb dissociation curve and respiratory gas transport in man. Resp. Physiol. *7:* 7, 1969.

166. HALDANE, J. S., and J. LORRAIN SMITH: The absorption of oxygen by the lungs. J. Physiol. *22:* 231, 1897.

167. DAWSON, T. J., and J. V. EVANS: Effect of hemoglobin type on the cardiorespiratory system of sheep. Amer. J. Physiol. *209:* 593, 1965.

168. BANCHERO, N., R. F. GROVER, and D. H. WILL: The pulmonary circulation in the llama, *Lama Glama.* Unpublished data.

169. TENNEY, S. M.: Respiratory control in chronic pulmonary emphysema; a compromise adaptation. J. Maine Med. Assoc. *48:* 375, 1957.

170. JONES, R. H., J. MACNAMARA, and E. A. GAENSLER: The effects of intermittent positive pressure breathing in simulated pulmonary obstruction. Amer. Rev. Resp. Dis *82:* 164, 1960.

171. OTIS, A. B.: Some physiological responses to chronic hypoxia. *In* Oxygen in the Animal Organism (Ed. by F. Dickens, and E. Neil). New York, Macmillan Co., 1964.

172. MOORE, F. D., J. H. LYONS, JR., E. C. PIERCE, JR., A. P. MORGAN, JR., P. A. DRINKER, J. D. MACARTHUR, and G. J. DAMMIN: Post-traumatic Pulmonary Insufficiency. Philadelphia, W. B. Saunders, 1969.

173. ARNDT, H., T. K. C. KING, and W. A. BRISCOE: Diffusing capacities and ventilation: perfusion ratios in patients with the clinical syndrome of alveolar capillary block. J. Clin. Invest. *49:* 408, 1970.

174. RAHN, H., and L. E. FARHI: Ventilation, perfusion and gas exchange. The $\dot{V}_A/\dot{Q}$ concept. *In* Handbook of Respiration, Washington, D.C., Amer. Physiol. Soc. Sec. 3, Vol. I, 1964, Ch. 30, p. 735.

175. WEST, J. B.: Ventilation/Blood Flow and Gas Exchange. Oxford, Blackwell Scientific Publications, 1965.

176. FISHMAN, A. P.: Respiratory gases in the regulation of the pulmonary circulation. Physiol. Rev. *41:* 214, 1961.

177. DALY, I. DEBURGH, and C. HEBB: Pulmonary and Bronchial Vascular Systems. London, Edw. Arnold, 1966.

178. READ, J., and J. LEE: Regional pulmonary vasoconstriction as an individual factor in the genesis of cor pulmonale, Amer. Rev. Resp. Dis. *96:* 1181, 1967.

179. GROVER, R. F., J. H. K. VOGEL, K. H. AVERILL, and S. G. BLOUNT, JR.: Pulmonary hypertension. Individual and species variability relative to vascular reactivity. Amer. Heart J. *66:* 1, 1963.

180. MEAD, J., and C. COLLIER: Relation of volume history of lungs to respiratory mechanics in anesthetized dogs. J. Appl. Physiol. *14:* 669, 1959.

181. PERMUTT, S., J. B. L. HOWELL, D. F. PROCTOR, and R. L. RILEY: Effect of lung inflation on static pressure volume characteristics of pulmonary vessels. J. Appl. Physiol. *16:* 64, 1961.

182. DuBois, A. B.: Resistance to breathing. *In* Handbook of Physiology, Washington, D.C., Amer. Physiol. Soc., Sec. 3, Respiration, Vol. I, 1964, Ch. 16, p. 451.

183. GRODINS, F. I.: Analysis of factors concerned in the regulation of breathing during exercise. Physiol. Rev. *30:* 220, 1950.

184. WEIL, J.V., E. BYRNE-QUINN, I. E. SODAL, W. O. FRIESEN, B. UNDERHILL, G. F. FILLEY, and R. F. GROVER: Hypoxic ventilatory drive in normal man. (In press) J. Clin. Invest. 1970.

185. ASTRAND, P-O., and K. RODAHL: Textbook of Work Physiology. New York, McGraw-Hill Book Co., 1970.

186. REEVES, J. T., R. F. GROVER, and J. E. COHN: Regulation of ventilation during exercise at 10,200 feet in athletes born at low altitude. J. Appl. Physiol. *22:* 546, 1967.

187. PETTY, T. L., D. B. BIGELOW, and B. E. LEVINE: The simplicity and safety of arterial puncture. J.A.M.A. *195:* 693, 1966.

188. SNIDER, G. L., and D. MALDONADO: Arterial blood gases in acutely ill patients. J.A.M.A. *204:* 991, 1968.

189. PAYNE, J. P., and D. W. HILL (Ed.): A symposium on oxygen measurements in blood and tissues and their significance. London, J & A Churchill, Ltd., 1966.

190. KREUZER, F.: Oxygen Pressure Recording in Gases, Fluids, and Tissues. Int. Symposium on O_2 Pressure Recording (Nijmegen). Basel, S. Karger, 1969.

191. CLARK, L. C., JR.: Monitor and control of blood and tissue oxygen tensions. Trans. Amer. Soc. Artif. Intern. Organs. *2:* 41, 1956.

192. BEIGELMAN, P. M., *et al.*: Severe diabetic ketoacidosis. J.A.M.A. *210:* 1082, 1969.

193. ALBERT, M. S., R. B. DELL, and R. W. WINTERS: Quantitative displacement of acid-base equilibrium in metabolic acidosis. Ann. Intern. Med. *66:* 312, 1967.

194. KASSIRER, J. P., P. M. BERKMAN, D. R. LAWRENZ, and W. B. SCHWARTZ: The critical role of chloride in correction of hypokalemic alkalosis in man. Amer. J. Med. *38:* 172, 1965.

195. ARBUS, G. S., L. A. HEBERT, P. R. LEVESQUE, B. E. ETSEN, and W. B. SCHWARTZ: Characterization and clinical application of the "significance band" for acute respiratory alkalosis. New Eng. J. Med. *280:* 117, 1969.

196. BRACKETT, N. C., JR., C. F. WINGO, O. MUREN, and J. T. SOLANO: Acid-base response to chronic hypercapnia in man. New Eng. J. Med. *280:* 124, 1969.

197. BRACKETT, N. C., JR., J. J. COHEN, and W. B. SCHWARTZ: Carbon dioxide titration curve of normal man. New Eng. J. Med. *272:* 6, 1965.

198. FLEMMA, R. J., and W. G. YOUNG: The metabolic effects of mechanical ventilation and respiratory alkalosis in postoperative patients. Surgery *56:* 36, 1964.

199. HALL, K. D., and V. VARTANIAN: Control of serum potassium levels in the hyperventilated postoperative cardiac patient. Southern Med. J. *61:* 416, 1968.

200. COLLIP, J. B., and P. L. BACKUS: The effect of prolonged hyperpnoea on the CO_2 combining power of the plasma, the carbon dioxide tension of alveolar air and the excretion of acid and basic phosphate and ammonia by the kidney. Amer. J. Physiol. *51:* 568, 1920.

201. RELMAN, A. S., B. ETSTEN, and W. B. SCHWARTZ: The regulation of renal bicarbonate reabsorption by plasma carbon dioxide tension. J. Clin. Invest. *32:* 972, 1953.

202. EICHENHOLZ, A., R. O. MULHAUSEN, W. E. ANDERSON, and F. M. MacDONALD: Primary hypocapnia: a cause of metabolic acidosis. J. Appl. Physiol. *17:* 283, 1962.

203. PLUM, F., and J. B. POSNER: Blood and cerebrospinal fluid lactate during hyperventilation. Amer. J. Physiol. *212:* 864, 1967.

204. SIESJÖ, B. K., and A. KJALLQUIST: A new theory for the regulation of the extracellular pH in the brain. Scand. J. Clin. Lab. Invest. *24:* 1, 1969.

205. GOFF, A. M., and E. A. GAENSLER: Hyperventilation syndrome. Respiration *26:* 359, 1969.

206. KILBURN, K. H.: Shock, seizures and coma with alkalosis during hyperventilation. Ann. Int. Med. *65:* 977, 1966.

207. BLOCK, A. J., and W. C. BALL, JR.: Acute respiratory failure. Ann. Int. Med. *65:* 957, 1966.

208. SEVERINGHAUS, J. W., F. N. HAMILTON, and S. COTEV: Carbonic acid production and the role of carbonic anhydrase in decarboxylation in brain. Biochem. J. *114:* 703, 1969.

209. MULHAUSEN, R., A. EICHENHOLZ, and A. BLUMENTALS: Acid-base disturbances in patients with cirrhosis of the liver. Medicine *46:* 185, 1967.

210. JAMES, I. M., *et al.*: Effect of induced metabolic alkalosis in hepatic encephalopathy. Lancet *2:* 1106, 1969.

211. SHEAR, L., H. L. BONKOWSKY, and G. J. GABUZDA: Renal tubular acidosis in cirrhosis. A determinant of susceptibility to recurrent hepatic precoma. New Eng. J. Med. *280:* 1, 1969.

212. LADÉ, R. I., and E. B. BROWN, JR.: Movement of potassium between muscle and blood in response to respiratory acidosis. Amer. J. Physiol. *204:* 761, 1963.

213. KILBURN, K. H.: Movements of potassium during acute respiratory acidosis and recovery. J. Appl. Physiol. *21:* 679, 1966a.

214. POLAK, A., G. D. HAYNIE, R. M. HAYS, and W. B. SCHWARTZ: Effects of chronic hypercapnia on electrolyte and acid-base equilibrium. J. Clin. Invest. *40:* 1223, 1961.

215. ROBIN, E. D.: Abnormalities of acid-base regulation in chronic pulmonary disease with special reference to hypercapnia and extracellular alkalosis. New Eng. J. Med. *268:* 917, 1963.

216. REFSUM, H. E.: Acid-base disturbances in chronic pulmonary disease. Ann. N.Y. Acad. Sci. *133:* 142, 1966.

217. RICHARDS, D. W., JR., and A. L. BARACH: The effects of oxygen treatment over long periods of time in patients with pulmonary fibrosis. Amer. Rev. Tube *26:* 253, 1932.

218. GAMBLE, J. L.: Chemical Anatomy, Physiology and Pathology of Extracellular Fluid. Boston, Dept. of Pediatrics, Harvard Medical School, 1941.

219. BLEICH, H. L.: Computer evaluation of acid-base disorders. J. Clin. Invest. *48:* 1689, 1969.

220. COHEN, M. L.: A computer program for the interpretation of blood-gas analysis. Computers and Biomedical Research *2:* 549, 1969.

221. CUNNINGHAM, D. J. C., J. M. PATRICK, and B. B. LLOYD: The respiratory response of man to hypoxia. *In* Oxygen in the Animal Organism. (Ed. by F. Dickens and E. Neil.) New York, Macmillan Co., 1964.

222. AVERY, W. G., P. SAMET, and M. A. SACKNER: The acidosis of pulmonary edema. Amer. J. Med. *48:* 320, 1970.

223. CAIN, S. M.: Increased oxygen uptake with passive hyperventilation in dogs. J. Appl. Physiol. *28:* 4, 1970.

224. COMROE, J. H., JR., E. R. BAHNSON, and E. O. COATES, JR.: Mental changes occurring in chronically anoxemic patients during oxygen therapy. J.A.M.A. *143:* 1044, 1950.

225. WESTLAKE, E. K., T. SIMPSON, and M. KAYE: Carbon dioxide narcosis in emphysema. Quart. J. Med. *24:* 155, 1955.

226. BATES, D. V., and R. V. CHRISTIE: Respiratory Function in Disease. Philadelphia, W. B. Saunders Co., 1964.

227. CAMPBELL, E. J. M.: Oxygen therapy in diseases of the chest. Brit. J. Dis. Chest *58:* 149, 1964.

228. BIGELOW, D. B., T. L. PETTY, B. L. LEVINE, G. F. FILLEY, and M. M. FINIGAN: The effect of oxygen breathing on arterial blood gases in patients with chronic airway obstruction living at 5200 feet. Amer. Rev. Resp. Dis. *96:* 28, 1967.

229. CHERNIACK, R. M., and K. HAKIMPOUR: The rational use of oxygen in respiratory insufficiency. J.A.M.A. *199:* 178, 1967.

230. GRAY, F. D., JR., and G. J. HORNER: Survival following extreme hypoxemia. J.A.M.A. *211:* 1815, 1970.

231. GERST, P. H., W. H. FLEMING, and J. R. MALM: Effect of acidosis and alkalosis upon the susceptibility of the heart to ventricular fibrillation. Circulation *30* (supp. 3): 83, 1964.

232. BURNS, J. W., J. C. ROSE, and P. A. KOT: The effects of acidosis on the tension-velocity relations in the isolated supported dog heart. Fed. Proc. *27:* 377, 1968.

233. CARESS, D. L., A. S. KISSACK, A. J. SLOVIN, and J. H. STUCKEY: The effect of respiratory and metabolic acidosis on myocardial contractility. J. Thorac. Cardiov. Surg. *56:* 571, 1968.

234. WILDENTHAL, K., D. S. MIERZWIAK, R. W. MYERS, and J. H. MITCHELL: Effects of acute lactic acidosis on left ventricular performance. Fed. Proc. *27:* 377, 1968.

235. COTEV, S., J. LEE, and J. W. SEVERINGHAUS: The effects of acetazolamide on cerebral blood flow and cerebral tissue P_{O_2}. Anesthesiology *29:* 471, 1968.

236. Turino, G. M., R. V. Lourenso, L. A. G. Davidson, and A. P. Fishman: The control of ventilation in patients with reduced pulmonary distensibility. Ann. N.Y. Acad. Sci. *109:* 932, 1963.

237. Holland, R. A. B.: Physiologic dead space in the Hamman-Rich syndrome. Amer. J. Med. *28:* 61, 1960.

238. West, J. B.: Effect of slope and shape of dissociation curve on pulmonary gas exchange. Resp. Physiol. *8:* 66, 1969.

239. Filley, G. F.: Pulmonary Insufficiency and Respiratory Failure. Philadelphia, Lea & Febiger, 1967.

240. Lord, G. P., C. Hiebert, and D. T. Francis: Postoperative hypoxia. New Eng. J. Med. *280:* 1300, 1969.

241. Salvatore, A. J., S. F. Sullivan, and E. M. Papper: Postoperative hypoventilation and hypoxemia in man after hyperventilation. New Eng. J. Med. *280:* 467, 1969.

242. Hedley-Whyte, H., H. Corning, M. B. Laver, W. G. Ansten, and H. H. Bendixen: Pulmonary ventilation-perfusion relations after heart valve replacement or repair in man. J. Clin. Invest. *44:* 406, 1965.

243. Prys-Roberts, C., G. R. Kelman, and R. Greenbaum: The influence of circulatory factors on arterial oxygenation during anesthesia in man. Anesthesia *22:* 257, 1967.

244. Ashbaugh, D. G., D. B. Bigelow, T. L. Petty, and B. E. Levine: Acute respiratory distress in adults. Lancet *2:* 319, 1967.

245. Ashbaugh, D. G., T. L. Petty, D. B. Bigelow, and T. M. Harris: Continuous positive-pressure breathing (CPPB) in adult respiratory distress syndrome. J. Thorac. Cardiov. Surg. *57:* 31, 1969.

246. McIntyre, R. W., A. K. Laws, and P. R. Ramachandran: Positive expiratory pressure plateau: improved gas exchange during mechanical ventilation. Canad. Anaesth. Soc. J. *16:* 477, 1969.

247. Bartels, H., E. Bücherl, C. W. Hertz, G. Rodewald, and M. Schwab: Methods in Pulmonary Physiology. New York, Hafner Publishing Co., Inc., 1963.

248. Michenfelder, J. D., W. S. Fowler, and R. A. Theye: CO_2 levels and pulmonary shunting in anesthetized man. J. Appl. Physiol. *21:* 1471, 1966.

249. Karis, J. H., M. H. Harmel, and B. F. Hoffman: Effects of pH and P_{CO_2} on the sensitivity of purkinje fibers to digitalis. Circulation *22:* 770, 1960.

250. Lyons, J. H., Jr., and F. D. Moore: Posttraumatic alkalosis: incidence and pathophysiology of alkalosis in surgery. Surgery *60:* 93, 1966.

251. Ayres, S. M., and W. J. Grace: Inappropriate ventilation and hypoxemia as causes of cardiac arrhythmias. Amer. J. Med. *46:* 495, 1969.

252. Borden, C. W., R. V. Ebert, and R. H. Wilson: Anoxia in myocardial infarction and indications for oxygen therapy. J.A.M.A. *148:* 1370, 1962.

253. McNicol, M. W., et al.: Pulmonary function in acute myocardial infarction. Brit. Med. J. *2:* 1270, 1965.

254. Valencia, A., and J. H. Burgess: Arterial hypoxemia following acute myocardial infarction. Circulation *40:* 641, 1969.

255. Foster, G. L., G. G. Casten, and T. J. Reeves: The effects of oxygen breathing in patients with acute myocardial infarction. Cardiovas. Res. *3:* 179, 1969.

256. CHERNIACK, R. M., and T. E. CUDDY: Respiratory insufficiency in acute myocardial infarction. Canad. Med. Assoc. J. *101:* 478, 1969.
257. NEILL, W. A.: Effects of arterial hypoxemia and hyperoxia on oxygen availability for myocardial metabolism. Amer. J. Cardiol. *24:* 166, 1969.
258. Editorial: Oxygen in acute myocardial infarction. Lancet *2:* 525, 1969.
259. DOWNING, S. E., N. S. TALNER, and T. H. GARDNER: Influences of arterial oxygen tension and pH on cardiac function in the newborn lamb. Amer. J. Physiol. *211:* 1203, 1966.
260. GELET, T. R., R. A. ALTSCHULD, and A. M. WEISSLER: Effects of acidosis on the performance and metabolism of the anoxic heart. Circulation *40* (Supp. 4): 60, 1969.
261. KATZ, A. M.: Contractile proteins of the heart. Physiol. Rev. *50:* 63, 1970.
262. WEIL, M. H., and H. SHUBIN (Editors): Diagnosis and Treatment of Shock. Baltimore, Williams & Wilkins Co., 1967.
263. LOWERY, B. D., C. J. CLOUTIER, and L. C. CAREY: Blood gas determinations in the severely wounded in hemorrhagic shock. Arch. Surg. *99:* 330, 1969.
264. CLOUTIER, C. T., B. D. LOWERY, and L. C. CAREY: Acid-base disturbances in hemorrhagic shock. Arch. Surg. *98:* 551, 1969.
265. BREDENBERG, C. E., *et al.:* Respiratory failure in shock. Ann. Surg. *169:* 392, 1969.
266. McLAUGHLIN, J. S., *et al.:* Pulmonary gas exchange in shock in humans. Ann. Surg. *169:* 42, 1969.
267. HALDANE, J. S.: Symptoms, causes, and prevention of anoxaemia. Brit. Med. J. *2:* 65, July 19, 1919a.
268. HALDANE, J. S., J. C. MEAKINS, and J. G. PRIESTLEY: The respiratory response to anoxaemia. J. Physiol. *52:* 420, 1919b.
269. LAMBERTSEN, C. J., P. HALL, H. WOLLMAN, and M. W. GOODMAN: Quantitative interactions of increased Po_2 and Pco_2 upon respiration in man. Ann. N.Y. Acad. Sci. *109:* 731, 1963.
270. BYRNE-QUINN, E., J. V. WEIL, I. E. SODAL, G. F. FILLEY, and R. F. GROVER: Decreased ventilatory response to hypoxia and hypercapnia in athletes. Physiologist *12:* 190, 1969.
271. NIELSEN, M., and H. SMITH: Studies on the regulation of respiration in acute hypoxia. Acta Physiol. Scand. *24:* 293, 1952.
272. PEPELKO, W. E., and S. M. CAIN: Effect of hypercapnia or propranolol on cooling rate and oxygen uptake of the dog. Physiologist *12:* 326, 1969.
273. CANZANELLI, A., M. GREENBLATT, G. A. ROGERS, and D. RAPPORT: The effect of pH changes on the *in vitro* O_2 consumption of tissues. Amer. J. Physiol. *127:* 290, 1939.
274. MONOD, J., J. P. CHANGEUX, and F. JACOB: Allosteric proteins and cellular controls systems. J. Molec. Biol. *6:* 306, 1963.
275. WYMAN, J., JR.: Linked functions and reciprocal effects in hemoglobin: A second look. Advances Protein Chem. *19:* 223, 1964.
276. ROSSI-BERNARDI, L., and F. J. W. ROUGHTON: The specific influence of carbon dioxide and carbamate compounds on the buffer power and Bohr effects in human haemoglobin solutions. J. Physiol. *189:* 1, 1967.

277. WYMAN, J., JR.: Heme proteins. Advances Protein Chem. *4:* 407, 1948.

278. BEAN, J. W.: Effects of oxygen at increased pressure. Physiol. Rev. *25:* 1, 1945.

279. DONALD, K. W.: Oxygen poisoning in man. Brit. Med. J. *1:* 667, 1947.

280. MARSHALL, J. R., and C. J. LAMBERTSEN: Interactions of increased Po_2 and Pco_2 effects in producing convulsions and death in mice. J. Appl. Physiol. *16:* 1, 1961.

281. VAN SLYKE, D. D.: Some points of acid-base history in physiology and medicine. Ann. N.Y. Acad. Sci. *133:* 5, 1966.

282. INGVAR, D. H., B. K. SIESJO, N. A. LASSEN, and E. SKINHOJ (Editors): Third international symposium on cerebral blood flow and cerebrospinal fluid. Scand. J. Clin. Lab. Invest. 22 (Supp. 102), 1968.

283. HARPER, A. M.: The interrelationship between aPco_2 and blood pressure in the regulation of blood flow through the cerebral cortex. Acta Neurol. Scand. *41* (Supp. 14): 94, 1965.

284. SEVERINGHAUS, J. W., R. A. MITCHELL, B. W. RICHARDSON, and M. M. SINGER: Respiratory control at high altitude suggesting active transport regulation of CSF pH. J. Appl. Physiol. *18:* 1155, 1963.

285. LEE, J. E., F. CHU, J. B. POSNER, and F. PLUM: Buffering capacity of cerebrospinal fluid in acute respiratory acidosis in dogs. Amer. J. Physiol. *217:* 1035, 1969.

286. LASSEN, N. A.: Brain extracellular pH: the main factor controlling cerebral blood flow. Scand. J. Clin. Lab. Invest. *22:* 247, 1968.

287. SHIMOJYO, S., O. M. REINMUTH, K. KOGURE, and P. SCHEINBERG: Cerebral blood flow and metabolism at different levels of hypoxia. Trans. Amer. Neurol. Assoc. *91:* 338, 1966.

288. HUCKABEE, W. E.: Relationships of pyruvate and lactate during anaerobic metabolism. I. Effects of infusion of pyruvate or glucose and of hyperventilation. J. Clin. Invest. *37:* 244, 1958.

289. SIESJO, B. K., L. GRANHOLM, and A. KJALLQUIST: Regulation of lactate and pyruvate levels in the CSE. *In* Ingvar, D. H., *et al.:* Third International Symposium on Cerebral Blood Flow and Cerebrospinal Fluid. Scand. J. Clin. Lab. Invest. 22 (Supp. 102): I:F, 1968.

290. SOLOWAY, M., W. NADEL, M. S. ALBIN, and R. J. WHITE: The effect of hyperventilation on subsequent cerebral infarction. Anesthesiology *29:* 975, 1968.

291. BATTISTINI, N., M. CASACCHIA, A. BARTOLINI, G. BAVA, and C. FIESCHI: Effects of hyperventilation on focal brain damage following middle cerebral artery occlusion. *In* Brock, M., *et al.:* Cerebral Blood Flow. Berlin, Springer Verlog, 1969, pp. 249–253.

292. GOTOH, F., J. S. MEYER, and Y. TAKAGI: Cerebral effects of hyperventilation in man. Arch. Neurol. *12:* 410, 1965.

293. REIVICH, M., P. J. COHEN, and L. GREENBAUM: Alterations in the electroencephalogram of awake man produced by hyperventilation: effects of 100% oxygen at three atmospheres (absolute) pressure. Neurology *16:* 304, 1966.

294. ROTHERAM, E. B., JR., P. SAFAR, and E. D. ROBIN: CNS disorder during mechanical ventilation in chronic pulmonary disease. J.A.M.A. *189:* 993, 1964.

295. LEUSEN, I., J. WEYNE, and G. DEMEESTER: Acid base and lactate/pyruvate changes in CSF and brain. *In* Third International Symposium on Cerebral Blood

Flow and Cerebrospinal Fluid (Ed. by D. H. Ingvar *et al.*). Scand. J. Clin. Lab. Invest. *22* (Supp. 102): I:G, 1968.

296. OLIVA, P. B.: Lactic acidosis. Amer. J. Med. *48:* 209, 1970.

297. CARRYER, H. M.: Tissue anoxia resulting from respiratory alkalosis. Proc. Staff Meet. Mayo Clin. *22:* 456, 1947.

298. TAKANO, N.: Blood lactate accumulation and its causative factors during passive hyperventilation in dogs. Jap. J. Physiol. *16:* 481, 1966.

299. BERRY, M. N., and J. Scheuer: Splanchnic lactic acid metabolism in hyperventilation, metabolic alkalosis and shock. Metabolism *16:* 537, 1967.

300. ZBOROWSKA-SLUIS, D. T., and J. B. DOSSETOR: Hyperlactatemia of hyperventilation. J. Appl. Physiol. *22:* 746, 1967.

301. HALPERIN, M. L., M. L. KARNOVSKY, and A. S. RELMAN: On the mechanism by which glycolysis responds to pH: implications of acid-base homeostasis. Clin. Res. *15:* 359, 1967.

302. ENGEL, K., P. A. Kildeberg, B. P. Fine, and R. W. Winters: Effects of acute respiratory acidosis on blood lactate concentration. Scand. J. Clin. Lab. Invest. *20:* 179, 1967.

303. CAIN, S. M.: Diminution of lactate rise during hypoxia by P_{CO_2} and β-adrenergic blockade. Amer. J. Physiol. *217:* 110, 1969.

304. SORBINI, C. A., V. GRASSI, E. SOLINAS, and G. MUIESAN: Arterial oxygen tension in relation to age in healthy subjects. Respiration *25:* 3, 1968.

305. WINTERS, R. W.: Terminology of acid-base disorders. Ann. N.Y. Acad. Sci. *133:* 211, 1966.

306. NAUNYN, B.: Der Diabetes Mellitus, 2nd ed. Vienna, A. Hölder, 1906.

307. VAN SLYKE, D. D., and G. E. CULLEN: Studies of acidosis. J. Biol. Chem. *30:* 289, 1917.

308. BARCROFT, J.: The Respiratory Function of the Blood. Part I. Lessons from High Altitude. Cambridge, Cambridge University Press, 1925, pp. 93, 99.

309. CAMPBELL, E. J. M.: Discussion of terminology of acid-base disorders. Ann. N.Y. Acad. Sci. *133:* 225, 1966.

310. VAN SLYKE, D. D.: Studies of acidosis. J. Biol. Chem. *48:* 153, 1921.

311. MICHAELIS, L.: Hydrogen Ion Concentration. Baltimore, Williams & Wilkins Co., 1926, pp. 94, 102.

312. KILDEBERG, P.: Personal communication, 1969.

313. KATZ, M. A.: pH vs. acid load. New Eng. J. Med. *280:* 1480, 1969.

314. PIERCE, E. C.: Further development of a simplified method for determining metabolic and respiratory pH factors. Trans. Amer. Soc. Artif. Intern. Organs *6:* 240, 1960.

315. JORGENSEN, K., and P. ASTRUP: Standard bicarbonate, its clinical significance and a new method for its determination. Scand. J. Clin. Lab. Invest. *9:* 122, 1957.

316. SIGGAARD-ANDERSEN, O: The pH-log P_{CO_2} blood acid-base nomogram revised. Scand. J. Clin. Lab. Invest. *14:* 598, 1962.

317. POSNER, J. B., and F. PLUM: Spinal-fluid pH and neurologic symptoms in systemic acidosis. New Eng. J. Med. *277:* 605, 1967a.

318. CAIN, S. M.: Personal communication, 1970.

Index of First Authors of Cited Literature

....

Subject Index